MW00843834

Climate Change Law

FOUNDATIONS OF ENVIRONMENTAL LAW

Environmental law plays a critical role in governing the way we use, preserve, and interact with the natural world. In a time of escalating environmental emergencies, this authoritative series will be pivotal in ensuring access to accessible, yet nuanced, overviews of the fundamental areas of law upon which the core frameworks and institutions of environmental law are based.

These rigorous and structured guides will offer an invaluable platform for exploring pressing issues in the field, from climate change and biodiversity, to indigenous rights and social responsibility. Titles in the series will be authored by leading scholars in the field, as well as a new generation of emerging experts, appealing to a broad and international readership of scholars, postgraduate students, lawyers and policy makers.

Climate Change Law
An Introduction

Karl S. Coplan

Professor of Law, Elisabeth Haub School of Law, Pace University, USA

Shelby D. Green

Professor of Law, Elisabeth Haub School of Law, Pace University, USA

Katrina Fischer Kuh

Professor of Law, Elisabeth Haub School of Law, Pace University, USA

Smita Narula

Haub Distinguished Professor of International Law, Elisabeth Haub School of Law, Pace University, USA

Karl R. Rábago

Principal, Rábago Energy LLC, USA

Radina Valova

Regulatory Vice President, Interstate Renewable Energy Council, USA

FOUNDATIONS OF ENVIRONMENTAL LAW

 Edward Elgar
PUBLISHING

Cheltenham, UK • Northampton, MA, USA

Published by
Edward Elgar Publishing Limited
The Lypiatts
15 Lansdown Road
Cheltenham
Glos GL50 2JA
UK

Edward Elgar Publishing, Inc.
William Pratt House
9 Dewey Court
Northampton
Massachusetts 01060
USA

A catalogue record for this book
is available from the British Library

Library of Congress Control Number: 2021947939

This book is available electronically in the **Elgar**online
Law subject collection
http://dx.doi.org/10.4337/9781839101304

Printed on elemental chlorine free (ECF)
recycled paper containing 30% Post-Consumer Waste

ISBN 978 1 83910 129 8 (cased)
ISBN 978 1 83910 130 4 (eBook)

Printed and bound in the USA

Contents

Figures

Acknowledgments

The authors thank Umair Saleem, Environmental Law Fellow for the Pace Environmental Law Program, for his work reviewing, editing, and coordinating the production of this book, and the Elisabeth Haub School of Law and Pace University for providing institutional support.

In addition, Pace Law students Audra Gale, Dana McClure, Cassie Jurenci, Taylor Keselica, and Claire McLeod provided valuable research assistance in support of this project.

Karl S. Coplan
Shelby D. Green
Katrina Fischer Kuh
Smita Narula
Karl R. Rábago
Radina Valova
December 2021

About the authors

Karl S. Coplan has been a Professor and Associate Professor of Law at Pace Law School and director of Pace Environmental Litigation Clinic since 1994. Prior to joining the Pace faculty, he practiced land use and environmental litigation for eight years with the New York City firm of Berle, Kass & Case. As the principal outside counsel for Riverkeeper, Inc., Professor Coplan and the Pace Environmental Litigation Clinic have brought numerous lawsuits enforcing the Clean Water Act and other environmental laws. In addition to directing the Pace Environmental Litigation Clinic, Professor Coplan teaches courses in Environmental Law and Constitutional Law.

Shelby D. Green is a Professor of Law at the Elisabeth Haub School of Law at Pace University, White Plains, NY. She is a graduate of Georgetown University Law Center. She teaches and writes in the areas of real estate transactions, property, housing, and historic preservation.

Katrina Fischer Kuh is the Haub Distinguished Professor of Environmental Law at the Elisabeth Haub School of Law at Pace University, where she teaches Administrative Law, Climate Change Law, Environmental Law, and Torts. She is co-editor of *The Law of Adaptation to Climate Change U.S. and International Aspects* (with Michael Gerrard) and publishes frequently on topics related to environmental law and climate change.

Smita Narula is the Haub Distinguished Professor of International Law at the Elisabeth Haub School of Law at Pace University, where she teaches International Environmental Law, Human Rights and the Environment, Environmental Justice, and Property. Smita has founded and directed numerous non-profit and higher education initiatives dedicated to the promotion and protection of human rights, and to social and environmental justice worldwide. She is former legal advisor to the UN Special Rapporteur on the Right to Food and publishes frequently on topics related to food systems and human rights.

Karl R. Rábago is the Principal of Rábago Energy, LLC, an energy regulatory consulting practice. He previously led the Pace Energy and Climate Center and taught Energy Law at the Elisabeth Haub School of Law at Pace University. Karl and has more than 30 years' experience in energy regulation, including as a public utility commissioner in Texas, a utility executive, a renewable energy developer, a non-profit advocate, and as a deputy assistant secretary with the US Department of Energy.

Radina Valova serves as Regulatory Vice President for the Interstate Renewable Energy Council, where she provides strategic direction and oversight of IREC's engagement in interconnection and grid modernization proceedings, and develops policy best practices for laying the technical and regulatory foundation for a 100% clean energy future. Prior to joining IREC, Radina served as Senior Staff Attorney for the Pace Energy and Climate Center, where she led the Center's engagement in New York's Reforming the Energy Vision, buildings and gas decarbonization, and clean and affordable energy for low- and moderate-income communities. Radina is a graduate of the Elisabeth Haub School of Law at Pace University, with J.D. Certificates in Environmental and International Law, and an LL.M. in Land Use and Sustainable Development.

Abbreviations

ACES	American Clean Energy and Security Act (2009)
ACHPR	African Commission on Human and Peoples' Rights
ACtHPR	African Court on Human and Peoples' Rights
AEP v. Connecticut	American Electric Power Co. v. Connecticut
AOSIS	Alliance of Small Island States
ASHRAE	American Society of Heating, Refrigerating and Air-Conditioning Engineers
AU	African Union
BACT	best available control technology
BAT	best available technology economically achievable
BCT	best conventional pollutant control technology
BDT	best demonstrated technology
BEPS	building energy performance standards
BINGOs	business and industry NGOs
BMPs	best management practices
BPT	best practicable control technology
BSER	best system of emissions reductions
BTRs	Biennial Transparency Reports
CAA	Clean Air Act (1970)
CEQ	Council on Environmental Quality
CEQA	California Environmental Quality Act (1970)
CESCR	Committee on Economic, Social and Cultural Rights
CHR	Commission on Human Rights
CO_2e	carbon dioxide equivalent
Comer	Comer v. Murphy Oil USA, Inc.

COP	Conference of the Parties
CRC	Convention on the Rights of the Child
CWA	Clean Water Act (1972)
DER	distributed energy resources
DG	distributed generators
DOJ	Department of Justice
ECHR	European Convention on Human Rights
ECtHR	European Court of Human Rights
EERS	Energy Efficiency Resource Standards
EIAs	environmental impact assessments
ENGOs	environmental NGOs
EPA	Environmental Protection Agency
ESG	environmental, social and governance
ETS	Emissions Trading Scheme
FERC	Federal Energy Regulatory Commission
FIP	federal implementation plan
GHGs	greenhouse gases
GLAC	Greater Los Angeles County
GSI	Green Stormwater Infrastructure
HVAC	heating, ventilation, and air conditioning
IACHR	Inter-American Commission on Human Rights
IACtHR	Inter-American Court of Human Rights
ICCPR	International Covenant on Civil and Political Rights (1966)
ICESCR	International Covenant on Economic, Social and Cultural Rights (1976)
IECC	International Energy Conservation Code
IGOs	Indigenous Peoples Organizations
IPCC	Intergovernmental Panel on Climate Change
Juliana	Juliana v. United States
Kivalina	Native Village of Kivalina v. ExxonMobil Corp.
Kyoto Protocol	Kyoto Protocol to UNFCCC (1997)
LAER	lowest achievable emissions rate

LED	light-emitting diode
LEED	Leadership in Energy & Environmental Design
LGMAs	local government and municipal authorities
MACT	maximum achievable control technology
MD&A	management's discussion and analysis
NAAQS	National Ambient Air Quality Standards
NDCs	nationally determined contributions
NEPA	National Environmental Policy Act
NGOs	nongovernmental organizations
NHTSA	National Highway Transportation Safety Administration
NIDIS	National Integrated Drought Information System
NSPS	New Source Performance Standards
NYCEEC	New York City Energy Efficiency Corporation
OAS	Organization of American States
OHCHR	Office of the High Commissioner for Human Rights
Paris Agreement	Paris Agreement to UNFCCC (2015)
PSCs	public service commissions
PUCs	public utility commissions
RACT	reasonably available control technology
RECs	Renewable Energy Certificates
REDD/REDD+	Reducing Emissions from Deforestation and Forest Degradation
RGGI	Regional Greenhouse Gas Initiative
RINGOs	research and independent NGOs
RPS	Renewable Portfolio Standards
SEC	Securities and Exchange Commission
SIDS	small island developing states
SIP	state implementation plan
TDR	transferable development rights
TFEU	Treaty on the Functioning of the European Union
TUNGOs	trade union NGOs
UDHR	Universal Declaration of Human Rights (1948)

UN	United Nations
UNDRIP	UN Declaration on the Rights of Indigenous Peoples
UNEP	UN Environment Programme
UNFCCC	United Nations Framework Convention on Climate Change (1992)
UPR	Universal Periodic Review
US	United States
VCEA	Virginia Clean Economy Act (2020)
VMT	vehicle miles traveled
WGC	women and gender constituency
YOUNGOs	youth NGOs
ZEV	zero-emission vehicles

Introduction to *Climate Change Law*

In the preceding decades, the attributes of climate change—including long time horizons, complex science, numerous and widespread contributors, and disparate geographic impacts—repeatedly intersected with political and legal processes to frustrate the development of effective mitigation laws and policies. Climate change thus lives up to its billing as a "super wicked" public policy problem defying ready solution.[1] Current commitments by countries to reduce atmospheric greenhouse gases (GHGs) fall far short of what is necessary to limit climate change and its impacts to tolerable levels. Even if every party honors its non-binding commitments under the Paris Agreement, models project that warming will, within this century, significantly exceed 2°C, the level of warming generally recognized as the maximum tolerable.[2] To put this in perspective, the present-day manifestations of "just" a little over 1°C warming—wildfires, flooding, storms—are proving to be deadly, costly, and dislocating; as discussed in the next chapter, scientists project even more serious, irreversible impacts with warming of even 1.5°C. Yet knowledgeable observers caution that "barring rapid global political, social, and technological transformations … we will be fortunate to limit temperature rise to 2.6°C, just as likely to reach 3.9°C, and the possibility of reaching 4.0°C or higher cannot be ignored."[3]

To avoid the devastating climate change impacts hovering at the edges of climate models, the present moment must be an inflection point marking a rapid transition from an extended period of study, debate, planning, and preliminary emission reduction efforts to a period of rapid, effective implementation of climate change mitigation measures. Myriad developments, ranging from the adoption of aggressive GHG emissions targets by some countries and some states within the United States (US) to the declining market value of

---

[1]   For further insight into why it is so difficult to develop climate change mitigation law and policy *see* Richard J. Lazarus, *Super Wicked Problems and Climate Change: Restraining the Present to Liberate the Future*, 94 CORNELL L. REV. 1153, 1159–1187 (2009).

[2]   UNITED NATIONS ENVIRONMENT PROGRAMME, EMISSIONS GAP REPORT 2019 Figure ES.4 (2019), https://www.unep.org/emissions-gap-report-2020.

[3]   J.B. Ruhl & Robin Kundis Craig, *4°Celsius*, 106 MINN. L. REV. (forthcoming 2021).

major fossil fuel companies, signal the possibility that climate change policy is turning this corner. But time is short. Humans emit about 40 Gt of carbon into the atmosphere per year, which puts us on track, by many estimates, to use up our carbon budget (the total amount of carbon that can be put into the atmosphere to retain a decent chance of keeping warming below 1.5°C or 2°C) by about the mid-2030s; in 2019, scientists warned that when the release of naturally sequestered carbon (from permafrost emissions or forest dieback, for example) is factored in, the world may have *already* used up the carbon budget.[4]

And the task to achieve adequate near-term carbon reductions is herculean. *Legal Pathways to Deep Decarbonization*, a comprehensive guide for how to achieve meaningful mitigation in the US, identifies over a thousand laws and policies relevant to decarbonization, providing some perspective on the scope and nature of what decarbonization will require.[5] Mitigation interventions will need to change not just how we generate energy (swapping out fossil fuel combustion for renewable sources of energy), but how we use it (swapping out cars with internal combustion engines for electric vehicles and natural gas appliances for electric ones). Adopting and implementing these interventions will, in turn, require broad participation, sustained political will, informed policy judgment, and sophisticated legal acumen. It is all hands on deck—indeed, the editors of *Legal Pathways to Deep Decarbonization* are coordinating an effort to engage lawyers pro bono across numerous fields of legal practice to contribute to decarbonization efforts.[6]

This underscores the importance for citizens, policymakers, and lawyers to be literate in the core aspects of climate change law and policy. Yet individuals interested to understand the contours of climate change law and policy face a challenge. The law and policy of climate change mitigation is overwhelming in its breadth, the diversity of its origins, and the complexity of its content. It arises and is enforced at all levels of government, from the adoption of energy-efficient building codes by local governments to federal laws governing the emissions of pollutants from industrial sectors. It is interpreted and applied by a diverse array of institutions, including dozens of federal and state agencies and federal and state courts in myriad jurisdictions. Relevant authorities include statutes, the regulations interpreting them, agency guidance and thousands of court decisions. And the underlying legal subjects with which climate change law and policy intersects include not just environmental

---

[4]   *Id.*

[5]   Legal Pathways to Deep Decarbonization in the United States (Gerrard & Dernbach eds.) (2018).

[6]   Model Laws for Deep Decarbonization in the United States, https://lpdd.org/pathways/ (last visited 4/21).

and energy law; nearly all legal subjects intersect in some significant manner with climate change. Indeed, at the Elisabeth Haub School of Law at Pace University, the once or current home institution of the authors of this book, faculty have discussed whether and how to include coverage of climate change *in every course taught at the school.*

The breadth and diversity of climate change law and the fact that it touches nearly every legal subject and practice area invite the question of whether it is productively considered a distinct area of law at all.[7] Yet, regardless of whether climate change law is ultimately understood to formally constitute a unique subject of legal inquiry, it is imperative—in the sense that our collective future depends upon it—for law to develop and support effective climate change mitigation policy.

Helping individuals seeking to engage with climate change law and policy to develop climate change law literacy is the *raison d'étre* of this book. It provides background information necessary for readers to knowledgably navigate climate change law and policy. The chapters scaffold understanding of key issues by introducing and defining key terms, authorities, and actors; identifying important policy considerations; explaining the contours of significant legal disputes and questions; and highlighting the most salient aspects of emerging mitigation policy and associated legal issues. Readers should develop a sense of the broad landscape of climate change law and policy as well as familiarity with important terms, institutions, actors, and issues that will equip them to independently take deeper dives into specific areas of interest. The goal, in short, is to make readers literate in climate change mitigation law and policy, thereby empowering them to engage further with climate change mitigation efforts.

The first chapter, "International Climate Change Treaty Regime," explains the key provisions of the Paris Agreement, emphasizing how and why domestic

---

[7]   *Compare* Frank H. Easterbrook, *Cyberspace and the Law of the Horse,* 1996 U. Chi. Legal F. 207 (1996) (arguing that "the best way to learn the law applicable to specialized endeavors is to study general rules") *with* Lawrence Lessig, *The Law of the Horse: What Cyberlaw Might Teach,* 113 Harv. L. Rev. 501, 503 (1999) (responding that the study of cyberlaw, and perhaps other specialized areas, may provide "lessons for law generally"). *See also* Douglas A. Kysar, *What Climate Change Can Do About Tort Law,* 42 Envtl. L. Rep. News & Analysis 10739 (2012) (explaining how climate change may influence the development and understanding of tort law); J.B. Ruhl & James Salzman, *Climate Change Meets the Law of the Horse,* 62 Duke L.J. 975, 1019 (2013) (evaluating whether the law of climate change adaptation constitutes a distinct field and observing that "it may very well be that no existing field of law is rendered obsolete by climate change, but that more than a Law of the Horse is needed—that is, a distinct field of climate adaptation theory and practice is nonetheless necessary and appropriate to manage policy goals that no individual field can address.").

policymaking will be crucial for achieving its objectives. Chapter 2, "Climate Law Primer: Mitigation approaches," then identifies and describes federal policies to reduce GHG emissions in the US, reviewing the basic structure of pollution control approaches and assessing existing and potential future application of these approaches to mitigation, including under the federal Clean Air Act. Chapter 3, "Introduction to Energy Law," focuses on the role of energy law in mitigation, providing an overview of public service and administrative law, and laws governing the development of energy projects, the regulation of energy in the buildings sector, and the transportation sectors. And Chapter 4, "Adaption to Climate Change at the Subnational Level," supplements the preceding analyses by highlighting the role of regional, state, and local law in developing and implementing climate change policy. Together, the first four chapters provide an accessible tour of current climate change policy in the US, with particular attention to identifying emerging issues and forecasting likely developments.

The remaining chapters consider important legal doctrines and ethical principles that inform, define, and shape the development of climate change policy. Chapter 5, "Litigating Government (In)Action on Climate Change," explores catalyst litigation to force governments to require stronger mitigation action through robust enforcement of existing statutes of general application or by invoking extra-statutory sources of government obligation, including common law and constitutional law. Chapter 6, "Human Rights and Climate Change," examines the relationship between human rights and climate change using the framework of international human rights law, including by considering how human rights law has evolved to elaborate states' obligations to mitigate and adapt to climate change, and to address the needs of those most vulnerable to climate-related harms. Chapter 7 focuses on possibilities for using law to prompt mitigation or compensation for climate harms by holding private actors that contribute to climate change accountable and/or persuading them to voluntarily reduce emissions. This includes evaluating lawsuits brought directly against large emitters and fossil fuel producers by those suffering harms from climate change and evaluating how disclosure under federal securities laws can support mitigation by private actors. The book concludes by turning the focus to individuals, observing that ultimately the development and implementation of effective climate change policy will rest on collective acknowledgment of a shared ethical responsibility. Chapter 8, "Why the Individual Ethics of GHG Emissions Matter to Climate Law," looks to prevailing ethical theories of utilitarianism, deontology, and virtue ethics to show that individuals have a moral responsibility to reduce and moderate their lifestyle greenhouse gas emissions.

1. International climate change treaty regime

Katrina Fischer Kuh

The chief product of over 25 years of international negotiations to address climate change, the Paris Agreement, can be summarized in one sentence: Most countries in the world have agreed to set their own targets for reducing emissions, report on their progress toward meeting those targets using agreed-upon metrics, and update those targets and make them more stringent as necessary over time. The procedural requirements involved (tracking and reporting on emissions, setting and submitting targets) are binding, but there are no binding substantive requirements in the sense that countries may choose their own targets and there is no formal penalty for failing to meet a target. Stripped of its terms of art—"nationally determined contributions," "global stocktakes"—the contours of the Paris Agreement reveal themselves to be rather simple and modest, belying the lengthy and difficult negotiations that produced them. That the core requirements of the Paris Agreement are simple and modest does not mean that they will not ultimately prove effective. The simplicity and modesty of the core provisions of the Paris Agreement does, however, evidence the continued and primary importance of individual country commitments and measures to mitigate climate change; signal that the Paris Agreement's implementation and evolution will be crucial to whether and to what extent it ultimately adds value to worldwide mitigation efforts; and invite the question of why it took 25 years to achieve an agreement of such limited scope.

This chapter will begin by explaining terms, concepts, and the role of institutional actors central to the international climate change treaty regime. It will then describe key components of the Paris Agreement, highlighting how it centers domestic mitigation laws and explaining why the Paris Agreement's ultimate contribution to mitigation cannot be predicted with confidence. The chapter will conclude by looking back and summarizing the international climate change negotiations and agreements that preceded and produced the Paris Agreement, emphasizing issues and developments that help to explain its provisions and bottom–up approach.

I   DEFINITIONS, CONCEPTS, AND ACTORS

Periodic assessment reports summarizing the science related to climate change published by the Intergovernmental Panel on Climate Change (IPCC), an entity formed in 1988 by the World Meteorological Organization and United Nations Environment Programme, guide countries in the negotiation and implementation of international agreements to address climate change.[1] IPCC reports engage the efforts of thousands of scientists around the world—the fifth assessment report involved 830 lead authors and review editors, over 1,000 contributors, and 2,000 expert reviewers from more than 80 countries—and are widely regarded as reliable and authoritative statements of scientific knowledge regarding the causes and impacts of climate change.[2] Reflecting the emerging nature and uncertainties of climate change science, assertions within IPCC reports are often accompanied by a characterization of the level of confidence for the assertion, which can range from very high confidence to very low confidence. Where an assessment of confidence is not possible, a description of the level of agreement and evidence that supports an assertion may be provided instead. In addition to detailed scientific and technical analysis, IPCC reports include a Summary for Policymakers which provides a more accessible overview of the most salient information in the report. Unlike full-length report chapters (which must be accepted by a plenary session that includes government representatives but which are not subject to editing by government representatives), the Summary for Policymakers is a consensus document subject to review and revision by governments and as a result tends to represent middle ground. Critiques of the IPCC process and the reports that it produces include that they do not adequately incorporate feasibility issues and understate worst-case risk.

The international climate treaty regime that the IPCC reports inform consists of three core agreements: the United Nations Framework Convention on Climate Change (UNFCCC) (1992),[3] Kyoto Protocol (1997),[4] and Paris

---

[1]   Reports of the IPCC are available from the IPCC's website at https://www.ipcc .ch/ (last visited March 1, 2021).

[2]   For an overview of the development and work of the IPCC, *see* UNION OF CONCERNED SCIENTISTS, THE IPCC: WHO ARE THEY AND WHY DO THEIR CLIMATE REPORTS MATTER? (July 2018), *available at* https://www.ucsusa.org/resources/ipcc -who-are-they (last visited March 2, 2021).

[3]   United Nations Framework Convention on Climate Change, May 9, 1992, S. Treaty Doc No. 102–38 (1992), 1771 U.N.T.S. 107 [hereinafter UNFCCC].

[4]   Kyoto Protocol to the United Nations Framework Convention on Climate Change, Dec. 11, 1997, 2303 U.N.T.S. 162 [hereinafter Kyoto Protocol].

Agreement (2015)[5]—the key provisions of which are described below. Decisions further elaborating on and implementing these agreements are made by a conference of the parties (COP). Over time, these COPs have evolved from government-centered negotiations to massive gatherings that include states, intergovernmental organizations, and a diverse array of non-governmental organizations (NGOs), climate change activists, academics, and representatives from the private sector. Recognized constituencies (groups of NGOs with diverse but broadly clustered interests or perspectives) include business and industry NGOs (BINGOs), environmental NGOs (ENGOs), farmers and agricultural NGOs (Farmers), indigenous peoples organizations (IGOs), local government and municipal authorities (LGMAs), research and independent NGOs (RINGOs), trade union NGOs (TUNGOs), the women and gender constituency (WGC), and youth NGOs (YOUNGOs). At modern COPs, formal proceedings and state-level negotiation are embedded in a rich milieu of workshops, side events, exhibits, and meetings.

The most significant tension in the international climate change negotiations is whether and how climate change agreements should reflect the relative responsibilities of countries in light of differences with respect to countries' contribution of emissions, relative development and wealth, and vulnerability to the impacts of climate change. Discerning those responsibilities is central to setting expectations with respect to the speed and extent of emission reductions by a country and the support to be provided or received by a country for mitigation and adaptation efforts. There are different perspectives for evaluating responsibility. With respect to thinking about relative responsibility vis à vis contribution to the problem (emissions), country emissions can be viewed through the lens of a country's historical or cumulative emissions, current total emissions, per capita emissions (i.e., emissions as compared to population), and through the lens of production-based accounting (emissions physically generated within a country, including producing goods that it then exports) versus consumption-based accounting of emissions (which charges the emissions associated with a good to the country where the good is ultimately consumed). To take one example, India has significant current emissions but has relatively low historical and per capita emissions and, because it exports carbon-intensive goods to other countries, lower emissions under a consumption-based accounting approach.[6] The charts shown in Figures 1.1

---

5   Paris Agreement to the United Nations Framework Convention on Climate Change, Dec. 13, 2015, *in* Rep. of the Conference of the Parties on the Twenty-First Session, U.N. Doc. FCCC/CP/2015/10/Add.1, annex (2016) [hereinafter Paris Agreement].

6   The different emission profiles of countries using these approaches can be found at: Hannah Ritchie, *Who has contributed most to global CO2 emissions?* OUR WORLD IN DATA (Oct. 1, 2019), https://ourworldindata.org/contributed-most-global-co2.

and 1.2 give a sense of the differences as between countries with respect to carbon dioxide emissions as between countries with respect to cumulative and annual carbon dioxide emissions.

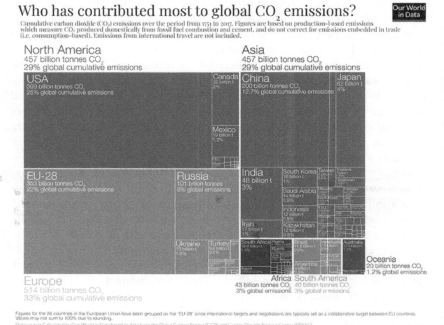

Figure 1.1   Who has contributed most to global CO2 emissions?

Hannah Ritchie, *Who has contributed most to global CO2 emissions?* OUR WORLD IN DATA (Oct. 1, 2019). Source: https://ourworldindata.org/contributed-most-global-co2.

Many countries in the Global North have high historical and current emissions, particularly when considered on a per capita basis, and the impacts that these countries face from climate change are, on the whole, projected to be later-occurring and less severe (at least initially). Many countries in the Global South have far lower historical and current contributions of emissions, particularly when considered on a per capita basis, but the impacts that these countries face from climate change are, on the whole, projected to occur sooner and be more severe. Countries in the Global North also tend to have, as a result of their wealth, greater capacity to adapt to the impacts of climate change than countries in the Global South. Additionally, historical emissions from the

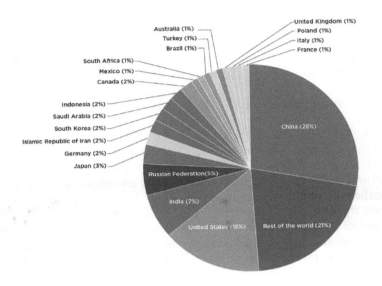

Figure 1.2   Each country's share of CO2 emissions

Each Country's Share of CO2 Emissions, UNION OF CONCERNED SCIENTISTS. Source: https://www.ucsusa.org/resources/each-countrys-share-co2-emissions (last updated Aug. 12, 2020).

Global North were chiefly produced during and as a result of industrialization; many countries in the Global South are still in the process of development and industrialization. As discussed in greater detail in Chapter 6, climate change thus presents a circumstance of markedly unequal burdens and unequal blame.

The international climate change regime has, from its inception, recognized both that the nature of climate change compels participation by all countries and that the Global North bears special responsibility and should take a leading role. Navigating these two principles to determine what, exactly, should be expected from different countries has, however, proved enormously difficult and the structure of the Paris Agreement reflects that difficulty.

II   OVERVIEW OF THE PARIS AGREEMENT

The Paris Agreement embodies the current state of international cooperation with respect to climate change. An overview of key components of the Paris Agreement follows. Importantly, the Paris Agreement does not include top–down binding requirements for parties to reduce emissions, instead leaving parties with the freedom to determine their own mitigation goals while mandating processes designed to nudge or persuade countries to achieve responsible levels of mitigation over time. This bottom–up structure centers domestic adoption and implementation of mitigation measures.

A   The Temperature Goal

The Paris Agreement announces as a key aim "[h]olding the increase in the global average temperature to well below 2°C above pre-industrial levels and pursuing efforts to limit the temperature increase to 1.5°C above pre-industrial levels."[7] The temperature goal provides an important benchmark for evaluating the sufficiency of global commitments to reduce emissions, allowing for comparison of the emissions reductions required to meet the temperature goal as compared to projected emissions. For over 15 years, international climate change negotiations proceeded without a clear temperature goal, indexing mitigation efforts to the somewhat nebulous aim of stabilizing greenhouse gas (GHG) concentrations at a level that would prevent dangerous anthropogenic interference with the climate system. In 2010, countries identified a long-term goal of limiting the global average temperature increase to below 2°C above pre-industrial levels; adoption of the more ambitious goal of limiting warming to well below 2°C and pursuing efforts to limit warming to 1.5°C is often identified as a signature achievement of the Paris Agreement.

Meeting the Paris Agreement's more aspirational 1.5°C temperature goal, however, seems increasingly out of reach. As of this writing and as discussed in greater detail below, the gap between the emissions reductions presently called for under the Paris Agreement and the reductions necessary to avoid even 2°C of warming is startlingly large, somewhere in the range of 15 gigatons of carbon dioxide equivalent by 2030.[8] Many knowledgeable observers concede that it is unrealistic to expect that the world will limit warming to

---

[7] Paris Agreement, *supra* note 5, art. 2, ¶1.
[8] UNITED NATIONS ENVIRONMENT PROGRAMME, EMISSIONS GAP REPORT 2019 xviii (November 2019).

anything close to 1.5°C, with much steeper increases in temperature likely.[9] The Paris Agreement's 1.5°C temperature goal may thus end up serving less as a meaningful goal and more as testament that mitigation efforts will not avoid significant climate change impacts, with devastating consequence.

The Paris Agreement's exhortation to limit warming to 1.5°C resulted in large measure from the advocacy of small island developing states (SIDS), organized as the Alliance of Small Island States (AOSIS), who are extremely vulnerable to climate change impacts.[10] These countries, in addition to being unusually vulnerable to climate change impacts (often because of geography compounded by lack of adaptive capacity), have also contributed diminishingly little to climate change. They are least to blame yet will be first (and perhaps most severely) to suffer, giving their voice outsized moral force and influence in the international climate change negotiations. Since 2008, AOSIS actively championed the more ambitious mitigation target of 1.5°C and supported the development of relevant scientific data as well as working to push that data into the international climate change negotiations. Employing the rallying cry "1.5°C to survive," AOSIS worked to persuade other countries, including other vulnerable countries and least developed countries, to support the 1.5°C temperature goal. Ultimately, a broad coalition of countries, including the "high ambition coalition" of developed and developing states, embraced the 1.5°C temperature goal during the negotiations and succeeded in enshrining it in the Paris Agreement.

In further support of the 1.5°C temperature goal, the decision adopting the Paris Agreement invited the IPCC to produce a special report on the impacts of global warming of 1.5°C above pre-industrial levels and related global GHG emission pathways.[11] The IPCC's report, *Global Warming of 1.5°C: An IPCC Special Report,* documents how warming of 2°C would generate significantly more dire impacts than warming of 1.5°C. Importantly, however, it *also* documents that even warming of 1.5°C would produce significant climate change impacts, particularly for SIDS, including impacts that could not be avoided through adaptation.[12] The report explains, for example, that while "[m]arine

---

[9]   J.B. Ruhl & Robin Kundis Craig, *4°Celsius*, 106 Minn. L. Rev. (forthcoming 2021).

[10]   For an overview of AOSIS and its crucial role in the adoption of a 1.5°C temperature goal in the Paris Agreement, *see* Lisa Benjamin & Adelle Thomas, *1.5 °C to stay alive? AOSIS and the long-term temperature goal in the Paris Agreement*, IUCN Academy of Environmental Law eJournal 7, (2016) 122–129.

[11]   Conference of the Parties on its twenty-first session, *Part two: Action taken by the Conference of the Parties at its twenty-first session* ¶ 21 FCCC/CP/2015/10/Add.1 (Jan. 29, 2016) [hereinafter Decision Adopting the Paris Agreement].

[12]   Intergovernmental Panel on Climate Change, *Global Warming of 1.5°C: An IPCC Special Report on the impacts of global warming of 1.5°C above pre-industrial*

systems and associated livelihoods in SIDS face higher risks at 2°C compared to 1.5°C," the impacts on coral reefs are extremely concerning at 1.5°C warming: "At 1.5°C, approximately 70–90% of global coral reefs are projected to be at risk of long-term degradation due to coral bleaching, with these values increasing to 99% at 2°C."[13]

The Paris Agreement's temperature goal is thus, in one sense, inadequate. Even if the world achieves its more aspirational temperature goal of limiting warming to 1.5°C, that will concededly not prevent dangerous anthropogenic interference with the climate system—it might, in fact, leave several atoll islands uninhabitable.[14] It is fair to critique the goal as regrettably underprotective. In another sense, however, the Paris Agreement's temperature goal is wildly ambitious. Limiting warming to 1.5°C, or even 2°C, requires emissions reductions and carbon removal on an unprecedented and massive scale over very short timeframes. It is also thus fair to critique the 1.5°C goal as unrealistic and unachievable. The tension between these perspectives for evaluating the temperature goal—the warming that must be avoided to prevent significant harm versus the feasibility of doing so—underscores the complexity of selecting a temperature goal, as well as the intractability of the climate change challenge.

B   Nationally Determined Contributions

The Paris Agreement requires parties to "prepare, communicate and maintain successive nationally determined contributions," (NDCs) or emissions reductions targets, and to "pursue domestic mitigation measures, with the aim of achieving the objectives of such contributions."[15] Parties determine their own emission reduction target; NDCs are recorded in a public registry maintained by the secretariat and must contain "information necessary for clarity, transparency and understanding."[16] Parties are required to update their NDCs every five years.

---

levels and related global greenhouse gas emission pathways, in the context of strengthening the global response to the threat of climate change, sustainable development, and efforts to eradicate poverty 175-312, SPM-29 (2018).

   [13]   *Id.* at 235.
   [14]   *Id.*
   [15]   Paris Agreement, *supra* note 5, art. 4(2).
   [16]   To this end, NDCs must conform to instructions on their preparation, accounting, and the like set forth in decision 1/CP.21 and any relevant decisions of the Conference of the Parties serving as the meeting of the Parties to the Paris Agreement. Collectively, these decisions are often referred to as the Paris Rulebook and provide detailed guidance on implementation of the Paris Agreement, including, importantly, detailing how Parties will report on their emissions and track progress toward imple-

These core provisions form a scaffolding for coordinating and encouraging international cooperation to achieve the Paris Agreement's temperature goal. Each party selects its own emissions target and there is no formal penalty for a party's failure to meet the emissions target identified in a previously submitted NDC. (Article 15 of the Paris Agreement establishes a mechanism to promote compliance consisting of a committee of experts but it is expressly facilitative in nature and stated to function in a non-adversarial and non-punitive manner.) The expectation and hope are that the act of complying with the required procedures of the Paris Agreement will—through the generation of reliable and shared information, reputational pressure, and financial support to developing countries—propel mitigation action by parties despite the lack of binding obligations to achieve specific emission reductions. The effectiveness of the Paris Agreement will turn on the extent to which the Paris Agreement's required processes—primarily the development and submission of NDCs—are implemented with rigor and produce transparency that encourages and disciplines countries to adopt and meet ambitious emissions targets proportionate to the reductions necessary to respect the temperature goal.

Some of the Paris Agreement's procedural requirements support ambitious mitigation action in ways that may not be immediately apparent but that could cause mandatory process to mature, over time, into obligations that become substantive. Although parties set their own emissions targets, the Paris Agreement specifies that "[e]ach Party's successive nationally determined contribution will represent a progression beyond the Party's then current nationally determined contribution and reflect its highest possible ambition."[17] While it is not yet clear what, exactly, it means for an NDC to be progressive and how it will be evaluated and enforced, robustly interpreted and applied, a requirement for progression could turn the procedural requirement to submit NDCs every five years into a meaningful substantive obligation as parties become required to reduce national emissions over time.[18]

Additionally, the Paris Agreement provides that parties' successive NDCs are to "be informed by the outcomes of the global stocktake."[19] The global stocktake refers to the process of "periodically tak[ing] stock of the implemen-

---

menting and achieving their NDC targets. Matters Relating to the Implementation of the Paris Agreement, Decs. 1/CP.24 and 3/CMA.1, UNFCC, U.N. Doc. FCCC/PA/CMA/2018/L.4 (2018) [hereinafter Paris Rulebook].

[17] Paris Agreement, *supra* note 5, art. 4(3).

[18] Lavanya Rajamani & Jutta Brunnée, *The Legality of Downgrading Nationally Determined Contributions under the Paris Agreement: Lessons from the US Disengagement*, 29 J. OF ENVTL. L. 537–551 (2017) (explaining that it would violate the spirit of the Paris Agreement to downgrade an NDC).

[19] *Id.* at art. 4(9).

tation of this Agreement to assess the collective progress towards achieving the purpose of this Agreement and its long-term goals."[20] The Paris Agreement requires that global stocktakes be conducted beginning in 2023 and every five years thereafter and states that "[t]he outcome of the global stocktake shall inform Parties in updating and enhancing, in a nationally determined manner, their actions and support in accordance with the relevant provisions of this Agreement, as well as in enhancing international cooperation for climate action."[21] As parties' NDCs must be informed by the outcome of the (presumably most recent) global stocktake, the global stocktake process should—if robustly implemented—push successive NDCs to become more progressive if a stocktake reveals a need for further mitigation.

Finally, these binding procedures—the submission of progressive NDCs responsive to global stocktakes—are in turn supported by yet another set of binding procedures related to their implementation, the enhanced transparency framework. Under the enhanced transparency framework, all parties are required to submit, in the form of biennial transparency reports (BTRs), national GHG inventory reports (anthropogenic emissions by sources and removals by sinks of GHGs) and information necessary to track progress made in implementing and achieving NDCs.[22] BTRs will be subject to a two-part review process that includes a technical expert review and a facilitative, multilateral consideration of progress.[23] The technical expert review is an objective assessment and is not meant to express a political judgment or otherwise comment on the appropriateness of a party's NDC or progress. The facilitative, multilateral consideration of progress will provide an opportunity for broader comment on a party's efforts, with parties allowed to submit written questions to other parties about their BTRs followed by a working group session (open to observers) where parties can again pose questions. The enhanced transparency framework procedures are thus designed to support the NDC and global stocktake procedures by ensuring the production of accurate and comparable information and furnishing a public platform for interrogating that information, hopefully thereby supporting the development of a strong international norm of ambitious mitigation effort.

---

[20]   *Id.* at art. 14.
[21]   *Id.*
[22]   Paris Agreement, *supra* note 5, art. 13(7).
[23]   *Id.* at art. 13(11).

C   Obligations of Conduct, Not Obligations of Result

The Paris Agreement combines binding requirements to engage in a specified process without compelling (in the form of top–down emission targets or penalties for failure to achieve an NDC) a substantive outcome. For this reason, it has been aptly described as "an internationally legally binding agreement, containing provisions with variable legal character."[24] Under the Paris Agreement, achieving emission reductions is not an obligatory commitment, but the hoped-for outcome as parties follow required procedures, norms develop around mitigation ambition, and parties are prompted to take voluntary mitigation action. The core provisions of the Paris Agreement are thus modest, in the sense that they create "obligations of conduct" as opposed to "obligations of result."[25]

The Paris Agreement's mix of mandatory process and non-binding substantive outcomes reflects a negotiating compromise crucial to achieving broad participation. Because there was not support for a new, international agreement on climate change in the US Senate, the United States would not have joined any agreement that imposed new substantive obligations (such as a binding emission reduction target). President Obama was able to enter the United States into the Paris Agreement through an executive agreement (which does not require ratification by the Senate) because US obligations under the Paris Agreement were already required under the UNFCCC or authorized under domestic law.

Allowing parties to set their own non-binding NDCs also achieved broad participation by forestalling the need to decide the relative responsibilities of different parties, including developed versus developing countries. One of the chief accomplishments of the Paris Agreement is to require the submission of NDCs by developing country parties, thereby engaging all countries, regardless of development status, in global mitigation efforts. The Paris Agreement achieves this alignment of developed and developing country effort by including developing countries in the Agreement's key processes (including most importantly by requiring developing countries to submit NDCs), while continuing to recognize and make accommodations for the different circumstances of those countries.

The Paris Agreement continues to recognize that developed countries should take the lead in mitigation and support mitigation and adaptation in developing countries. The Agreement repeatedly describes higher expectations for devel-

---

[24]   Jacob Werksman, *Remarks on the International Legal Character of the Paris Agreement*, 34 MD. J. INT'L L. 343, 353 (2019).

[25]   *Id.* at 361 (quotation and citation omitted).

oped country parties. For example, the Agreement explicitly recognizes that peaking of GHG emissions "will take longer for developing country Parties"[26] and exhorts that developed country Parties "should continue taking the lead by undertaking economy-wide absolute emission reduction targets," while "[d]eveloping country Parties should continue enhancing their mitigation efforts, and are encouraged to move over time towards economy-wide emission reduction or limitation targets in the light of different national circumstances."[27] The Paris Agreement also states that NDCs will reflect a party's "common but differentiated responsibilities and respective capabilities, in the light of different national circumstances"[28] and at myriad junctures observes that the extent of support received from developed country Parties will be considered in evaluating developing country ambition and implementation. The Agreement also seeks to strengthen commitments by developed countries to provide financial, technology-transfer, and capacity-building support to developing countries by requiring that developed country Parties report information on the support that they have provided during the BTRs, subject to technical expert review, as part of the enhanced transparency framework.[29]

D   Loss and Damage, Forestry, Adaptation, Internationally Transferred Mitigation Outcomes, and Market Mechanisms

The submission of NDCs, the enhanced transparency framework, and the global stocktakes constitute the central structure of the Paris Agreement. Many other important issues are referenced essentially as a placeholder either to continue existing work on the issue or flagging the issue as one for future negotiation. In some instances, the Agreement commits to continue existing efforts (as with respect to addressing loss and damage, the preservation and management of forests as carbon sinks, and adaptation) and in others it identifies an issue but defers important implementation decisions to future negotiations (as with respect to the use of internationally transferred mitigation outcomes and market mechanisms).

The phrase "loss and damage" refers to harms from climate change that exceed adaptive capacity and therefore cannot be prevented or fixed. Whether and how the international climate change treaty regime should deal with loss and damage, in particular when suffered by countries that have contributed relatively little to climate change, remains a highly disputed issue. Vulnerable

---

[26]   Paris Agreement, *supra* note 5, art. 4(1).
[27]   *Id.* at art. 4(4).
[28]   *Id.* at art. 4(3).
[29]   *Id.* at art. 13(9).

countries are keen for redress of their climate change harms while large emitters, primarily in the Global North, are wary of potential liability for claims of loss and damage. The Paris Agreement continues preexisting international efforts to address loss and damage (for example, through insurance and risk transfer) and situates them within the Paris Agreement. Article 8 provides that the Warsaw International Mechanism for Loss and Damage is subject to the authority and guidance of the Paris Agreement's COP and exhorts parties to "enhance understanding, action and support ... on a cooperative and facilitative basis with respect to loss and damage."[30] Reflecting developed country skittishness about potential liability for loss and damage, the decision adopting the Paris Agreement admonishes, however, that Article 8 "does not involve or provide a basis for any liability or compensation."[31]

The management of forests to maximize carbon sequestration has also long been a focus of the international climate change treaty regime. The Paris Agreement encourages Parties to conserve and enhance sinks and reservoirs of GHGs, primarily forests, through existing frameworks such as the Warsaw Framework for REDD+.[32] The Warsaw Framework for REDD+ was adopted at COP19 in December 2013 and provides methodological and financing guidance for the implementation of REDD+ activities, which include reducing emissions from deforestation and forest degradation, the conservation of forest carbon stocks, sustainable management of forests, and the enhancement of forest carbon stocks. The Paris Agreement thus blesses continued work through existing processes to promote the conservation and enhancement of sinks and reservoirs, most notably forests, without materially changing the treaty regime's approach to the issue.

The Paris Agreement's approach to adaptation (or preparation for the impacts of climate change as opposed to efforts to reduce climate change (mitigation)) likewise builds on prior efforts, while also incorporating new and potentially powerful transparency related to adaptation. The agreement announces a global goal on adaptation[33] and encourages parties to strengthen their cooperation on enhancing action on adaptation, taking into account the existing Cancun Adaptation Framework.[34] The Paris Agreement requires that parties engage "as appropriate" in adaptation planning processes and imple-

---

[30]   *Id.* at art. 8.

[31]   Decision 1/CP.21, *supra* note 11, art. 5.

[32]   Paris Agreement, *supra* note 5, art. 5.

[33]   *Id.* at art. 7(1) (announcing the global goal of "enhancing adaptive capacity, strengthening resilience and reducing vulnerability to climate change, with a view to contributing to sustainable development and ensuring an adequate adaptation response in the context of the temperature goal").

[34]   *Id.* at art. 7(7).

mentation and provides that parties should submit and periodically update adaptation communications describing adaptation progress and needs.[35] Global stocktakes will evaluate adaptation, including by reviewing the adequacy and effectiveness of adaptation and also evaluating the support provided for adaptation and reviewing the overall progress made in achieving the global goal on adaptation.[36]

With respect to loss and damage, forestry, and adaptation, the Paris Agreement creates continuity, largely recommitting to the continuation and strengthening of existing initiatives. With respect to another important issue central to the operation of the Paris Agreement, the question of how Parties can use emission reductions achieved through non-domestic mitigation to fulfill their own NDCs and the operation of an international carbon market, the Paris Agreement essentially serves as a placeholder, deferring difficult decisions to future negotiations.

Article 6(2) of the Paris Agreement approves, in concept, the idea that Parties can use internationally transferred mitigation outcomes to achieve NDCs and clarifies that such arrangements are to be voluntary, authorized by participating Parties, and conform to some general guidelines (for example, use robust and transparent accounting and avoid double counting).[37] Article 6(4) contemplates the establishment of an international carbon market and creates a mechanism, to be overseen by a new body designated by the COP, which will develop rules, modalities, and procedures for its implementation. Article 6(4) instructs that one aim of the market shall be to achieve an overall mitigation in global emissions, directs that a share of proceeds from its operation be put toward administrative expenses and supporting adaptation in vulnerable developing countries, and specifies that emissions reductions should not be counted toward achievement of both the host Party's NDC and another Party's NDC. The decision adopting the Paris Agreement further states that emissions reductions created by Article 6(4) must be "additional to any that would otherwise occur," referencing the idea that proceeding with business as usual should not have produced the reductions.[38]

But the Paris Agreement leaves decisions about crucial details for the implementation of both country-to-country exchanges and the operation of an international carbon market, including what will count as an internationally transferred mitigation outcome and how additionality will be evaluated, to be negotiated at future meetings of the COP. Ultimately, the country-to-country

---

[35] *Id.* at art. 7(9)–(12).
[36] *Id.* at art. 7(14).
[37] Paris Agreement, *supra* note 5, at art. 6 (2).
[38] Decision Adopting the Paris Agreement, *supra* note 11, at ¶ 38.

exchanges of mitigation outcomes and a functional international carbon market will likely be a key component for the achievement of NDCs, particularly as ambition increases, underscoring the central importance of the work yet to be done.

III EVOLUTION OF THE CLIMATE CHANGE TREATY REGIME

The Paris Agreement embodies the modern climate change treaty regime. Some understanding of the evolution of the climate change treaty regime nonetheless provides important context for understanding the Paris Agreement, including most notably its marriage of non-binding substantive elements and binding procedures.

The Paris Agreement is a legal instrument negotiated under the UNFCCC. The UNFCCC establishes the general system of governance that anchors the climate change treaty regime and, like the Paris Agreement, achieved universal participation. The UNFCCC can perhaps best be understood as an international agreement to agree to solve a common problem, climate change. The UNFCCC states the broad goal of "stabilization of greenhouse gas concentrations in the atmosphere at a level that would prevent dangerous anthropogenic interference with the climate system," creates a process for parties to submit national emission inventories, and requires developed country parties identified in Annex I of the Agreement to submit national mitigation plans and emission inventories.[39]

The UNFCCC also articulates principles to guide international cooperation to address climate change.[40] One key principle in the UNFCC is the idea that countries have common but differentiated responsibilities with respect to climate change. The concept captures the idea that all countries should work to solve climate change, but that their relative contributions to that effort must acknowledge differences in terms of the volume of emissions they have contributed to the problem and their development and capacities. The concept is particularly salient for thinking about the relative obligations of the Global North and Global South.

The UNFCCC's articulation of common but differentiated responsibilities is of continuing import because subsequent negotiations have failed to produce agreement on the meaning of the concept in practice (as applied, for example, to determine the appropriate mitigation obligations of different countries). Indeed, the practical meaning of common but differentiated responsibilities

---

[39] UNFCCC, *supra* note 3, art. 2, 4.
[40] *Id.* at art. 3.

(sometimes referred to as differentiation) has proved to be the stickiest wicket in development of international cooperation on climate change. The UNFCCC asserts that parties should act "in accordance with their common but differentiated responsibilities and respective capabilities."[41] It goes on to assert that "the developed country Parties should take the lead in combating climate change and the adverse effects thereof,"[42] having explained in the preamble that "the largest share of historical and current global emissions of GHGs has originated in developed countries, that per capita emissions in developing countries are still relatively low and that the share of global emissions originating in developing countries will grow to meet their social and development needs."[43] The agreement further emphasizes the development prerogative, stating that "Parties have a right to ... sustainable development" and that "Parties should cooperate to promote a supportive and open international economic system that would lead to sustainable economic growth and development in all Parties, particularly developing country Parties."[44] Applying this principle to assign obligations for emission reductions has proved fraught.

As discussed above, by allowing countries to set their own substantive emission targets, the Paris Agreement avoids the need to agree upon the relative responsibility of parties for mitigation. This reflects, in part, the great difficulty of translating the concept of common but differentiated responsibilities into practice. Indeed, disputes about the nature of common but differentiated responsibilities are key to understanding not only the structure of the Paris Agreement, but also why the agreement that immediately preceded it—the Kyoto Protocol—proved to be a dead end.

Negotiations immediately following the UNFCCC produced the Kyoto Protocol which set top–down binding GHG emission targets, termed quantified emission limitation and reduction objectives, which were typically around 5% below 1990 levels for the first compliance period (2008–2012) and 18% below 1990 levels in the second compliance period (2013 to 2020).[45] These

---

[41]   *Id.* at art. 3.

[42]   *Id.* at art. 3.

[43]   Preamble to the UNFCCC, *supra* note 3.

[44]   UNFCCC, *supra* note 3, art. 3.

[45]   That the world initially adopted this top–down approach to controlling GHG emissions likely reflects, in part, a reflexive effort to replicate the success of the world's approach to controlling the emission of ozone-depleting substances as embodied in the Vienna Convention for the Protection of the Ozone Layer (which, like the UNFCCC, set forth a very general agreement to agree to reduce the emission of ozone-depleting substances) and the Montreal Protocol on Substances That Deplete the Ozone Layer (which, like the Kyoto Protocol, imposed top–down binding emission limits). Notably, some substitutes for the ozone-depleting substances being phased out under the Montreal Protocol are also potent GHGs. The Montreal Protocol has since been

targets were only imposed on developed country parties identified in Annex I to the UNFCCC who ratified the Kyoto Protocol. The Kyoto Protocol established a cap and trade regime that incorporated flexibility mechanisms through which Annex I parties could meet their emissions targets by purchasing emission credits through International Emissions Trading (Article 17), supporting Joint Implementation projects in other developed countries to obtain Emission Reduction Units (Article 6), or obtain Certified Emission Reductions by supporting projects in developing countries through the Clean Development Mechanism (Article 12).[46]

That the initial interpretation of the UNFCCC's common but differentiated responsibilities imposed no emission targets on developing countries ultimately became the downfall of the Kyoto Protocol's top–down approach. The Kyoto Protocol never achieved universal participation. The US never ratified the Kyoto Protocol (primarily out of concern that no emission reductions targets were required from developing countries), Canada withdrew, some important Annex I countries did not submit new emission targets in the Kyoto Protocol's second compliance period (Japan, New Zealand, Russia), and developing country parties were never subject to emission targets. Moreover, many have called into question the volume of emission reductions ultimately achieved by the Kyoto Protocol.

The climate change treaty regime has since evolved into the bottom–up approach of the Paris Agreement described above. The significance of the Kyoto Protocol to international climate change mitigation efforts is debated. One area of learning under the Kyoto Protocol that will likely prove important is how the experience implementing the Clean Development Mechanism under the Kyoto Protocol—which has been subject to great criticism—will support and inform the development of rules relating to carbon markets under Article 6(4) of the Paris Agreement. And one issue presently being negotiated is whether and how to merge the Kyoto Protocol, in particular the potential migration of projects under the Clean Development Mechanism and Joint Implementation, into the structure of the Paris Agreement.

CONCLUSION

The takeaway from the current status of the international climate change treaty regime is both hopeful and worrisome. It has taken an exhausting nearly 30-year process to get us to what is, in some ways, a new starting

---

amended, through the Kigali Amendment, to limit the reliance on those GHG-heavy substitutes.

[46] Kyoto Protocol, *supra* note 4, at arts. 6, 12, 17.

point in the form of the Paris Agreement. And, as described above, it will take significant energy, focus, and commitment to successfully implement the Paris Agreement. What we must hope is that this time we are at the bottom of a ladder with more rungs, one that, if we climb, it will lead to effective international cooperation on mitigation.

2. Climate law primer: mitigation approaches

Karl S. Coplan

I   INTRODUCTION AND OVERVIEW

Climate "mitigation" refers to measures taken to reduce greenhouse gas (GHG) emissions with the goal of decreasing, or "mitigating" the severity of anthropogenic global warming. This chapter will provide a basic overview of legal mechanisms to regulate and limit environmental pollution and consider their application to reducing GHG emissions in the United States.[1] The Intergovernmental Panel on Climate Change (IPCC) has established GHG emissions budgets through mid-century for limiting global warming to 1.5°C, or to 2.0°C. These emissions targets are based on the levels of cumulative anthropogenic GHG emissions determined by climate models to result in an unacceptable risk of exceeding these temperature limits. The IPCC budgets are based on a two-thirds chance of meeting the target temperature limit if the target emissions level is reached.

Based on the SR15 report issued in October 2018, the IPCC emissions budget for limiting global warming to 1.5°C was 420 Gt of CO_2e. Staying on track to achieve this goal would require an approximate 45% reduction in global GHG emissions from current levels by 2030, and net zero emissions by 2050. Achievement of the IPCC global emissions budget will take coordinated global action. However, as explained in greater detail in Chapter 1, individual nations will make nationally determined contributions (NDCs) to the mitigation of global emissions, and national environmental laws and regulations will be needed to implement these national mitigation goals.

---

[1]   In keeping with the introductory nature of this book, the analysis in this chapter is highly simplified, and makes no claim to be an in-depth treatment of the subject. For a deep dive into regulatory measures to achieve GHG reductions in the US, *see* Legal Pathways to Deep Decarbonization in the United States (M. Gerrard & J. Dernbach eds., Environmental Law Institute 2019).

Fifty years of environmental law development in the United States illustrate the basic approaches to limiting emissions of ecologically harmful pollutants. These approaches include mandatory emissions controls on the one hand, and economic incentives for pollution reduction on the other. Mandatory emissions controls may be established on an industry-wide basis based on available control technologies or may be determined source-by-source as needed to meet environmental quality goals. Economic incentives for pollution reductions include taxing harmful emissions, subsidies for low-pollution technology, and caps on a category of emissions with tradeable emissions allowances. The earliest comprehensive environmental regulatory programs, the 1970 Clean Air Act (CAA) and the 1972 Clean Water Act (CWA), relied primarily on the first approach, which established emissions controls on a source-by-source basis. Industry resistance to this approach, pejoratively referred to as "command and control," led to more flexible trading-based approaches in later regulatory provisions, such as the Acid Rain Trading Program incorporated into Title IV of the CAA in the 1990 Amendments to the CAA.

Although climate change is unique among environmental problems in its scope and in the ubiquity of the emissions that cause it, GHG emissions are a form of emissions into the air. Thus, some GHG-emitting activities are already regulated under the CAA. These regulations include mileage standards for new motor vehicles as well as GHG emissions rates for fossil fuel power plants. Other activities not currently subject to regulation are potentially subject to future regulation under existing statutory authority. Future specific climate legislation seems certain eventually.

Policy analyses of feasible pathways to achieve the IPCC mitigation goals contemplate that both emissions limits and economic incentives will be necessary if the United States is to meet these goals.[2] As of this date, the United States does not have any comprehensive regulatory scheme designed to implement the necessary economy-wide reductions in GHG emissions. This chapter will accordingly review the basic structure of pollution control approaches, examples of each approach, an assessment of the success of each approach in achieving its goals, and the potential future application of these basic environmental law approaches to the mitigation of GHG emissions in the United States. This chapter will also review the history of regulation of GHG under the existing statutory authority of the CAA, and the potential for further regulation under existing CAA authority.

---

[2]   The most comprehensive analysis of policy measures needed to achieve the IPCC goals is Legal Pathways to Deep Decarbonization (Michael B. Gerrard & John C. Dernbach eds., Environmental Law Institute 2019).

II   EMISSIONS CONTROL MANDATES

A   Structure

Mandatory emissions controls are the most direct form of pollution control and were among the first forms of substantive federal environmental regulation adopted by Congress in the 1970s. The concept of mandatory controls is quite simple, even if their implementation becomes complicated. Environmental contamination sources are identified and are subject to numerical limits on the volumes of pollution they emit. Examples of such limits include point source water pollution controls implemented under the CWA as well as controls on air pollution sources under the CAA. Implementation of these pollution controls requires standards for formulating the appropriate level of emissions, procedures for calculating and applying these standards to individual sources, and means for monitoring compliance and enforcing against non-compliance. In general, mandatory pollution controls are implemented by requiring permits for specified sources of pollution and incorporating specific limits into individual facility permits. Federal and state agencies may share implementation authority in an example of cooperative federalism. Under such a system of environmental federalism, the Environmental Protection Agency (EPA) sets minimum national standards, while states are charged with implementing these standards through administration of permit requirements and environmental planning. Although principles of state sovereignty preclude Congress from requiring states to participate in a national program of environmental regulation, it can incentivize state participation by providing federal funding for state implementation, and for the EPA to step in and administer the program in those states that fail to.[3]

A system of mandatory emissions controls includes several structural elements. First, there must be a technical basis for the formulation of quantitative emissions limits. Second, there must be a system for administering these source-specific limits, usually in the form of a system of individual permits. Third, there must be a mechanism for monitoring compliance with emissions limits and enforcing compliance by those pollution sources that are out of compliance.

Specific pollutes → smoke stack vs ambient

---

[3]   *See* New York v. United States, 505 U.S. 144 (1992); Hodel v. Virginia Surface Mining Control Board, 452 U.S. 264 (1981).

B   Formulation of Limits

In order to impose quantifiable emissions limits on pollution sources, there must be some basis for determining the appropriate permissible level of emissions for each source. Specific emissions limits are generally either based on achieving environmental quality goals or based on the emissions reductions available from implementation of emissions control technologies. These two methods differ in both the underlying regulatory philosophy and the complexities of administration.

1   Environmental quality-based limits

Environmental quality-based limits are simple in principle: Pollution from all sources must meet limits set at a level that assures achievement of the desired level of environmental quality. As of this date, the United States does not have any comprehensive regulatory scheme designed to implement the necessary economy-wide reductions in GHG emissions to the level necessary to ensure an acceptable level of environmental quality. Basing pollution limits on avoiding adverse ecological harm is most directly tied to improving environmental quality and is perceived to avoid wasting resources on unneeded treatment or allowing insufficient treatment where more is environmentally necessary. On the other hand, the underlying principle—that of allowing pollution so long as the air or water that receives it can absorb it without unacceptable harm—can be criticized as making dilution the solution to pollution. The cost of pollution control, however, is not usually a consideration in establishing environmental quality-based pollution limits.

While simple in principle, administration of an environmental quality-based pollution control system is quite complex in practice. At a minimum, it requires four steps. First, there must be an identification and quantification of the ecological parameters that constitute acceptable environmental quality for a given resource, whether it be a lake, river, estuary, or the ambient air. Second, there must be a determination of where these minimum environmental quality criteria are not being met. Third, there must be a determination of the amount and location of pollutant reduction needed to bring the environmental resource into compliance with the standard. Fourth, there must be an allocation of the necessary pollutant reductions among the sources of relevant pollutants.

Each of these steps is fraught with technical and policy complications. Determination of the appropriate environmental quality criteria requires a policy judgment about the appropriate uses of the environmental resource in question—such as whether a given stretch of river should be suitable for drinking water, or swimming, or supporting a trout fishery, or just for transportation purposes. Determining the appropriate criteria also implicates combined science and policy questions in assessing what level of environmental contamination poses an unacceptable risk to the desired use of the resource. Identification of areas not meeting standards involves field deployment of environmental monitoring to measure compliance with the parameters during different seasons and weather and climate conditions. Determining the amount of pollution reduction needed in those ecosystems out of compliance with the standards requires sophisticated identification and modeling of all natural and anthropogenic sources of pollution entering the system and the complex interactions between pollutants, and between the natural environment and the pollution sources. And, finally, allocation of the needed emissions reductions involves the regulatory agency in the politically fraught enterprise of determining which individual sources should bear the additional costs of achieving the desired level of environmental quality, and which sources should not bear these costs.

Tech that's widely available

2 Technology-based limits

An alternative to environmental quality-based pollution limits is to base limits on available pollution control technologies. The underlying philosophy of technology-based limits is that all environmental pollution is wrong and should be avoided to the extent that cost-effective control technologies are available. The level of pollution control thus turns on the control technologies available and does not turn on the existence of proven environmental harm flowing from a particular source, or group of sources, of pollution.

Implementation of technology-based pollution standards usually requires a determination of the appropriate level of pollution reduction for a given category of pollution sources. Control technologies are usually established for categories and subcategories of industries, based on an assessment of the cost and effectiveness of pollution controls in use in a given industry. Determination of the appropriate control technology may turn on an assessment of the cost of controls and the environmental benefits achieved, with the relative weighting of the cost factor suggested by the statutory definition of the level of technology. Statutory technology-based standards established by Congress under the CAA range from "reasonably available control technology" to "maximum achievable control technology." While the level of pollution control mandated may be based on the selection of a particular control technology, the resulting standard is usually stated as a performance standard that does not mandate the use of that particular technology. Mandating a level of performance, rather than mandating a particular technology, allows and encourages the development of more cost-effective technologies that meet the standard.

Although technology-based pollution controls are usually based on an assessment of existing pollution control technologies, there may also be "technology forcing" standards, which set performance standards more stringent than available with proven technologies. Industry is thus forced to develop technologies capable of meeting the forward-looking standards, or face penalties for noncompliance.

Because technology-based standards are established by industry category rather than based on the unique environmental conditions of individual ecosystems across the country, one set of standards can be applied nationwide. The determination of the available pollution control technologies and industry costs demands considerable administrative resources and technical expertise, but may be less daunting an administrative task than determining environmental variables for each ecosystem. Nationwide standards also have the advantage of providing a level playing field, so that industries in every state and region have equivalent pollution control obligations.

C   Incorporation Into Permits

Whether the underlying pollution control standards are based on achieving desired environmental quality results or based on available pollution control technologies, specific standards must usually be developed and applied for each significant pollution source. This is true even for nationally applicable technology-based performance standards, as the exact quantities of permissible pollutant discharges will depend on the capacity and production levels of each pollution source. A pollution permit requirement allows regulatory agencies to spell out the exact pollution discharge limits that apply to each source, as well as to provide for monitoring and reporting requirements.

Permits may be administered by the federal government (usually the EPA) or by state environmental agencies. The three primary pollution control programs in the United States—CAA, CWA, and the Resource Conservation and Recovery Act—each contemplate a form of environmental federalism in which the EPA establishes national standards and guidelines, but the state environmental agencies have the option to administer the permitting program subject to EPA oversight. The EPA administers permits in those states that opt out of permit administration.

Although most fixed source controls under the CAA and the CWA are source-specific, and incorporated into permits, it is worth noting that pollution control mandates need not necessarily be individualized through permits issued source by source. Pollution controls may also be implemented by industry category. Motor vehicles emissions standards established under CAA § 202 are an example of such a control. Nor do all pollution control requirements take the form of numerical emissions limits; environmental regulation may take the form of requiring specific technologies or practices, such as best management practices (BMPs) required for industrial and municipal stormwater pollution control.[4] Agencies may also grant "general" or "nationwide" permits that may be invoked instead of an individual permit for those pollution sources in the covered categories.

D   Compliance and Enforcement

Effective pollution control requires a system of compliance monitoring and enforcement. In order to motivate pollution sources to comply with regulatory limits, they must have an expectation that noncompliance will be discovered and that the sanctions for noncompliance will be more onerous than the avoided cost of compliance.

Regulatory pollution control systems incorporate compliance monitoring both through agency inspections and self-monitoring and reporting by regulated facilities. Discharge and emissions sampling and testing requirements are ordinarily included in the terms of permits for individual sources. In addition, permits may provide for a right of inspection and emissions sampling by the environmental agencies.

Pollution control requirements are subject to a broad array of enforcement tools under the overlapping system of federal and state administration. At the federal level, the EPA has the authority to issue administrative compliance orders and assess penalties for violations. The US Department of Justice (DOJ) can seek civil penalties and orders in federal court on behalf of the EPA. Statutory civil penalties of up to $57,317 per day per violation (depending on the statute) are possible. For violations that satisfy the requisite *mens rea* requirements, the DOJ can seek more substantial criminal penalties, including substantial jail terms. State-level enforcement options generally mirror those available to the federal government. In addition, the pollution control laws in the United States generally authorize citizen enforcement in federal court for injunctive relief and the assessment of penalties for violations and an award of attorneys' fees for the cost of bringing the enforcement suit.

E   Examples of Pollution Control Mandates

The CAA and the CWA provide good examples of the experience with pollution control mandates. Each of these statutes combines both environmental-quality-based and technology-based standards. The CAA's National Ambient Air Quality Standards program is considered as an example of a quality-based emissions control program, and the CWA's National

---

[4]   *See generally* Michael L. Clar, P.E. et al., Stormwater Best Management Practice Design Guide: Volume 1 General Considerations 38 (2004).

Pollutant Discharge Elimination System permitting program is considered as an example of technology-based standards. The CAA's hydrocarbon emissions standards for motor vehicles provides an example of technology forcing standards.

F　　　Clean Air Act National Ambient Air Quality Standards

Title I of the CAA establishes an ambitious program for setting health and welfare-based air quality standards and mandates the achievement of these standards through a system of air pollution controls. The standards are known as the National Ambient Air Quality Standards, or NAAQS. The NAAQS program is implemented jointly between the EPA and the states. First, the EPA establishes a list of "criteria" pollutants meant to identify those pollutants that threaten public health or welfare. The current list of criteria pollutants includes sulfur oxides, nitrogen oxides, carbon monoxide, ozone, and lead. Next, the EPA establishes primary NAAQS at levels that, "allowing an adequate margin of safety, are requisite to protect the public health."[5] This environmental quality standard is based entirely on protecting public health; the EPA does not consider the measures or costs that would be required to meet the standard in setting the standard.[6] States are then charged with the task of developing a state implementation plan, or SIP, demonstrating how the state will achieve and maintain compliance with the NAAQS throughout the state through pollution control and management measures. The CAA initially required states to bring areas out of attainment into compliance with the NAAQS within five years, extendable to no more than twelve years[7]—however, these deadlines were extended in 1990 for areas with intractable air pollution problems.[8] The EPA reviews SIP submissions, and may disapprove a SIP that fails to meet statutory and regulatory requirements. If a state fails to remedy defects in its plan as identified by an EPA disapproval, the EPA may adopt its own federal implementation plan, or FIP, allowing the EPA to directly regulate pollution sources and transportation systems that contribute to the violation of air quality standards within the state. In addition, the EPA is authorized to limit federal highway funding to a state that fails to remedy defects in its SIP.

The NAAQS program, in combination with tailpipe automobile emissions standards adopted under CAA Title II, achieved huge progress in improving the nation's air quality. However, 50 years after enactment, several regions of

---

[5]　Clean Air Act (C.A.A.) § 109(b)(1), 42 U.S.C. § 7409(b)(1).

[6]　*See* Whitman v. American Trucking Associations, Inc., 531 U.S. 457 (2001).

[7]　C.A.A. § 172(a)(2), 42 U.S.C. § 7502(a)(2).

[8]　C.A.A. § 181, 42 U.S.C. § 7511.

the United States remain out of compliance with air quality standards for ozone and particulate matter.

G   Clean Water Act Technology-based Standards

The Federal Water Pollution Control Act Amendments of 1972 incorporated a phased-in approach to technology-based water pollution standards, with a goal of achieving a zero-pollutant discharge standard by 1985.[9] The initial technology-based pollution limitation goal was called best practicable control technology (BPT) and was required to be achieved by July 1, 1977. Municipal sewage treatment plants were subject to a similar standard reflecting secondary treatment technology. A more stringent standard, best available technology economically achievable (BAT), was to be achieved by all dischargers (other than municipal sewage treatment plants) by July 1, 1983. All dischargers were required to have an individual permit for their discharge, and the permitting program was called the National Pollutant Discharge Elimination System in order to reflect the Congressional goal of achieving zero pollutant discharge.[10] In the CWA Amendments of 1977, Congress abandoned the requirement that all dischargers achieve BAT for all pollutants, and adopted a new, intermediate standard of discharge for non-toxic pollutants, called best conventional pollution control technology (BCT).[11] New sources of water pollution were subject to a different set of standards, called New Source Performance Standards (NSPS).[12]

The CWA tasked the EPA with formulating the BPT, BCT, BAT, and NSPS that would apply to each industry category and subcategory. In determining the appropriate available treatment technology, the EPA was directed to consider a variety of factors, including the cost of implementing the treatment technology.[13] The CWA's inclusion of cost considerations varied for each level of technology, and avoided specifying a strict cost–benefit analysis for the reductions achieved by each industry subcategory. For the initial BPT standard, the statute specified consideration of "the total cost of application of technology in relation to the effluent reduction benefits to be achieved from such application."[14] For the more stringent BAT standard, the EPA was simply directed to

---

[9]   Federal Water Pollution Control Act Amendments of 1972, Pub. L. No. 92-500, §§ 2, 101(a)(1), 301(b)(1), 86 Stat. 816, 844 (1972).

[10]   Clean Water Act (C.W.A.) § 402, 33 U.S.C. § 1342 (2012).

[11]   Clean Water Act of 1977, Pub. L. No. 95-217, § 42, 91 Stat. 1582 (1977).

[12]   C.W.A. § 306, 33 U.S.C. § 1316 (2012).

[13]   C.W.A. § 304(b), 33 U.S.C. § 1314(b) (2012).

[14]   C.W.A. § 304(b)(1)(B), 33 U.S.C. § 1314(b)(1)(B) (2012).

consider "the cost of achieving such effluent reduction."[15] For the intermediate BCT standard, the statute directs a more complex cost-effectiveness evaluation of pollution control technologies.[16]

The EPA has interpreted the statutory direction to weight costs against the effluent reduction benefits for the BPT standard to base standards on the "average of the best" pollution control technology in use in an industry so long as the overall industry cost is not "wholly disproportionate" to the overall water quality benefit, even if particular applications of the technology at particular locations might not result in a water quality benefit in excess of its cost of implementation.[17] The EPA interprets the BAT cost standard to require implementation of the "best of the best" technology in use so long as the industry as a whole can afford the cost of compliance (even if individual plants might not be able to).[18]

Despite the EPA missing the deadlines for initial implementation of BPT, the success of the BPT standard in improving water quality led Congress to reduce the application of the stricter BAT standards to toxic and non-conventional pollutants in the 1977 CWA Amendments. This effectively abandoned the 1972 goal of achieving a zero-water pollution discharge standard.

H   Clean Air Act Technology-forcing Standards

One of the few examples of a technology-forcing standard is the hydrocarbon reduction standard written into the 1970 Clean Air Act, which required a 90% reduction in the emissions of hydrocarbons and carbon monoxide from light duty vehicles manufactured for model year 1975, as compared to model year 1970.[19] At the time of enactment, no emissions control technology capable of meeting this limit was in use. The adoption of this technology-forcing standard led to industry development of catalytic converter technology. Although this technology was ultimately universally implemented, it was only after several extensions of the compliance date.[20]

---

[15]   C.W.A. § 304(b)(2)(B), 33 U.S.C. § 1314(b)(2)(B) (2012).
[16]   C.W.A. § 304(b)(4)(B), 33 U.S.C. § 1314(b)(4)(B) (2012).
[17]   *See* Rybachek v. U.S. E.P.A., 904 F.2d 1276, 1289 (9th Cir. 1990) (costs not "wholly disproportionate"); Kennecott Copper Corp. v. E.P.A., 612 F.2d 1232, 1238 (10th Cir.1979) ("average of the best").
[18]   *See Rybachek*, 904 F.2d at 1290; American Meat Institute v. E.P.A., 526 F.2d 442, 463 (7th Cir. 1975) ("best performer in an industry category").
[19]   Clean Air Act of 1970, Pub. L. No. 91-604 §§ 6(a), 202(b)(1)(a), 84 Stat. 1690 (1970).
[20]   *See* International Harvester v. Ruckelshaus, 478 F.2d 615 (D.C. Cir. 1973); C.A.A. § 202(b)(1), 42 U.S.C. § 7521(b)(1).

I   Assessment of Application to Greenhouse Gas Emissions

Any one of these emissions control approaches, or a combination of them, could at least in theory be applied to control GHG emissions within the United States to meet the nation's hypothetical commitment to achieve GHG emissions reductions consistent with the IPCC's emissions pathways aimed at limiting global warming to 1.5 or 2.0° Celsius. Application of existing CAA authority to develop NAAQS and technology-based limits will be considered in Section IV of this chapter. As discussed below, the EPA has already implemented mileage standards for new automobiles in order to mitigate vehicle GHG emissions, and also issued regulations establishing technology-based performance standards for GHG emissions from new power plants. With appropriate authority, the EPA might take a NAAQS-like approach to achieving national reductions in GHG emissions, with the actual reduction mechanisms to be determined and implemented on a state-by-state basis through implementation plans. Similarly, with new authority, the EPA might adopt industry-wide technology-based standards for existing sources. Since technology-based standards are not designed to achieve a specific level of overall emissions reduction, adoption of technology-based standards by themselves will not guarantee achievement of economy-wide emissions reductions targets. It is worth noting that despite the incorporation of both technology and quality standards in both the CWA and the CAA, after five decades of implementation, these programs have not fully achieved their stated goals of achieving healthy air and fishable and swimmable water throughout the United States. In both cases, state-level implementation of the federal standards and incomplete federal oversight have fallen short.[21]

III   MARKET-BASED AND ECONOMIC INCENTIVES FOR POLLUTION REDUCTION

Direct regulation of pollution sources has been criticized as being too inflexible and economically inefficient, as it may require the same level of pollution

---

[21]   *See generally* Regulatory Impact Analysis of the Final Revisions to the National Ambient Air Quality Standards for Ground-Level Ozone, United States Environmental Protection Agency 1 – 16 (2015); *Impaired Waters and TMDLs Program in your EPA Region, State or Tribal Land,* United States Environmental Protection Agency (Sep. 13, 2018), https://www.epa.gov/tmdl/impaired-waters -and-tmdls-program-your-epa-region-state-or-tribal-land. *See also Waters Assessed as Impaired due to Nutrient-Related Causes,* United States Environmental Protection Agency (Sep. 3, 2020), https://www.epa.gov/nutrient-policy-data/waters-assessed -impaired-due-nutrient-related-causes.

control at different sources without regard for the differential costs of achieving that level of pollution reduction. Economic theorists posit that overall welfare would be improved if greater reductions were implemented at sources with the cheapest cost of pollution reduction and lesser reductions were required where the pollution was more expensive to control. Tradeable permit systems, known as "cap and trade," as well as pollution fees allow pollution sources to decide for themselves whether it is in their economic interest to pay the fee (or pay to acquire an emissions allowance) or to expend the necessary funds to reduce pollution from the source. Cap and trade, and pollution fees, thus enlist market and economic forces to achieve voluntary economy-wide pollution reductions. Both of these pollution pricing schemes provide an economic incentive to develop and implement new pollution reduction technologies, while allowing market forces to determine the most cost-effective technology. Subsidies for low-pollution activities provide a similar market incentive to reduce pollution-emitting activities. Each of these emissions control strategies can potentially be applied to GHG emissions, and GHG emissions trading systems have already been adopted in several US regions as well as in the European Union.

A   Cap and Trade

One way to enlist the market and economic incentives to control pollution is a system of tradeable emissions rights. This is generally known as "cap and trade," as the system of regulation imposes a system-wide cap of emissions, divides the capped amount into tradeable pollution permits, then allows pollution sources to buy and sell these permits as needed to cover their remaining emissions after implementing any pollution control measures that are less expensive than the tradeable permits. The price for a given amount of pollution will vary according to the supply and demand for the permits. The more stringent the cap is compared to the level of uncontrolled emissions, the lower the supply of emissions permits, which would lead to a higher relative price. On the other hand, the price will also be limited by the cost of pollution control at the source with the lowest cost of pollution abatement, as that source will choose to reduce their pollution and sell permits instead of paying the higher price for a permit.

1   Structure of a cap-and-trade system
Not all environmental pollution problems are susceptible to a system of tradeable emissions rights. First and foremost, a trading system only achieves its environmental protection goals if the pollutant involved is fungible across the geographic region subject to trading. Pollutants that cause highly localized environmental impacts, or "hot spots," are not appropriate for trading because

the purchaser of excess emissions rights will be able to continue to cause local pollution hot spots that are not mitigated by corresponding reductions by another pollution source somewhere else. In addition, there must be some mechanism to determine the level of the overall cap, ideally based on the environmental quality goal sought to be achieved.

In addition to a fungible pollutant, an emissions trading scheme requires some mechanism for distributing pollution permits, a market for trading in permits, and a system for compliance monitoring and enforcement. Some cap-and-trade programs include market-calming provisions to create additional allowances if the market price exceeds a set level. Permits can be auctioned by the government, or they can be distributed to existing pollution sources in proportion to their historical emissions, or by some other formula. Public commodities markets can provide a forum for trading of permits. Environmental agencies can perform the monitoring and enforcement function in the same way as for pollution control mandates.

2   Examples of pollutant trading systems

To date, the most comprehensive and successful pollutant trading system is the acid rain pollution trading program established by the 1990 Amendments to the CAA and incorporated into Title IV of the Act. The pollution trading system applied to fossil fueled electric generating units throughout the United States, and set a declining emissions cap for sulfur dioxide that, when fully phased in, was approximately 50% of the national emissions level before the cap. Emissions permits were distributed proportionately to existing power plants subject to the cap, based on their fuel source and generating capacity. The program was successful in meeting its emissions reductions goals and achieved the emissions reductions at a cost that was 57% lower than the estimated cost of pollution controls without the emissions trading system.[22]

B   Application of Cap and Trade to Greenhouse Gas Reductions

Following the success of the Title IV Acid Rain trading program, some international and regional US GHG trading systems have been established. In implementing the Kyoto Protocol to the United Nations Framework Convention on Climate Change, the European Union instituted a trading system for GHG emissions within the EU. The Emissions Trading Scheme (ETS) applies to

---

[22]   *See* A. Denny Ellerman, Paul L. Joskow & David Harrison, Jr., Emissions Trading in the United States: Experience, Lessons and Considerations for Greenhouse Gases, Pew Center on Global Climate Change Report, 15 tbl.2 (2003).

electricity generators and specified industrial facilities.[23] In the US, several eastern states set up the Regional Greenhouse Gas Initiative (RGGI),[24] which sets a declining cap on emissions from electricity generation in participating states. RGGI has been successful in substantially reducing emissions in the ten participating states, with a 45% reduction in covered emissions for 2020 as compared to a 2005 baseline.[25] The ETS was less successful in its early implementation, as the program allowed for too many emissions allowances, but the third phase, in effect since 2013, seeks to reduce emissions by 1.74% per year.[26]

Because of the ubiquitous penetration of GHG penetration into all aspects of the global economy and daily life in the developed world, a cap-and-trade system is widely viewed as an attractive policy tool for implementing needed emissions reductions. The great advantage of cap and trade as an emissions control strategy is that it provides certainty about the level of emissions reductions to be achieved, since the cap can be set commensurate with national commitments to achieve a specific percentage reduction in GHG emissions. A cap-and-trade system for GHG emissions would allow market forces to facilitate emissions reductions by those actors in the economy who could most easily (and cheaply) reduce their emissions, while allowing those economic sectors with the greatest emissions reductions challenges to purchase emissions allowances while they transition to cleaner energy and processes as the cap declines and the price of allowances increases. The Kyoto Protocol included emissions trading as an essential element of its international emissions reduction agreement. In addition, the 2009 American Clean Energy and Security Act (ACES), the only comprehensive climate legislation to pass at least one house of Congress, relied on a national cap-and-trade program to achieve its emissions reductions targets. ACES failed in the Senate and it never became law.

Cap and trade has both conservative and progressive critics. Conservatives assert that as the allowances decline, the market price for allowances will be driven so high that the program would be a substantial drain on the economy. Some estimates of the likely future price of emissions allowances to achieve necessary emissions reductions would be as much as $1,000 per ton of CO_2e,

---

[23]   *See generally* EU Climate Change Policy: The Challenge of the New Regulatory Initiatives (M. Peeters & K. Deketelaere eds., Edward Edgar Publishing Limited 2006).
[24]   *See* The Regional Greenhouse Gas Initiative, www.rggi.org (last visited Mar. 5, 2020).
[25]   Chios Carmody, *A Guide to Emissions Trading under the Western Climate Initiative*, 43 Canada-US L.J. 148, 161 (2019).
[26]   *Id.* at 163.

or the equivalent of a $10 per gallon tax on gasoline.[27] Progressives worry that the higher prices for GHG-intensive commodities, including food, gasoline, and heating fuel, would disproportionately affect lower-income households. Emissions trading mechanisms may also result in the concentration of harmful co-pollutant emissions in environmental justice communities where older emitting facilities are located.[28] In addition, the 2008 mortgage crisis and market collapse has resulted in a distrust of markets as a mechanism to allocate values and obligations in society. In addition, an economy-wide GHG emissions cap, including non-CO_2 emissions from agriculture, would be very complex to administer and enforce.

C   Emissions Taxes

Emissions taxes work similarly to tradeable emissions rights in that they also put a price on continued emissions. As with a cap-and-trade system, pollution sources with a low cost of pollution control will pay to reduce their emissions rather than pay the tax, as long as the tax is more expensive. Pollution sources with high control costs will pay the tax and continue their emissions rate. In this way, as with tradeable emissions rights, an emissions tax encourages the most economically efficient pollution controls, as, on an economy-wide basis, the most cost-effective pollution controls will be implemented.

Economists generally prefer pollution taxes to other forms of environmental control, as an emissions tax provides certainty about the price to be paid per unit of pollution, and the impact of the pollution reduction on the overall economy.[29] In theory, the tax rate for a given level of pollution reduction should be the same whether the reduction is achieved through cap and trade or a tax; the difference is that a cap-and-trade system provides certainty about the level of reductions, while a tax provides certainty about the price. The level of pollution reduction to be achieved by a given tax rate may be predicted through economic modeling but is uncertain. A pollution tax can, in theory, also result in the most economically efficient level of pollution reduction to be implemented

---

[27]   Joeri Rogelj, et al., Mitigation pathways compatible with 1.5°C in the context of sustainable development 152, in Special Report on Global Warming of 1.5°C 15, Intergovernmental Panel on Climate Change (2018), https://www.ipcc.ch/site/assets/uploads/sites/2/2018/07/SR15_SPM_version_stand_alone_LR.pdf.

[28]   *See* Alice Kaswan, *Environmental Justice and Domestic Climate Change Policy*, 38 ENVTL. L. REP. NEWS & ANALYSIS 10287 (2008).

[29]   *See* Roberta F. Mann, *The Case for the Carbon Tax: How to Overcome Politics and Find Our Green Destiny*, 38 Envt'l L. Rep. 10118, 10120 (2009); Jonathan B. Wiener, *Global Environmental Regulations: Instrument Choice in Legal Context*, 108 Yale L.J. 677, 682 (1999).

if the tax rate is set equal to the value of the presumed harm from each unit of pollution subject to the tax. However, this economic efficiency goal is hard to achieve in practice, as the dollar value of environmental harms from a unit of pollution are difficult to calculate, and pure economic efficiency calculations ignore regressive distributional impacts of pollution harms.

D   Examples of Pollution Taxes

There are few examples in existing law of pollution taxes adopted as a comprehensive approach to address an environmental problem. The closest analogies to a pollution tax are waste disposal fees, such as landfill tipping fees based on tons of waste disposed, and municipal sewage treatment fees imposed on industrial sources based on loadings of specified pollutants such as biological oxygen demand. These fees have the economic effect of incentivizing waste reduction. Beverage container deposit laws might also be seen as a form of environmental tax—those who litter or otherwise fail to return the container pay the tax, while those who reduce their litter and recycle the container receive their deposit back.

E   Application of Pollution Taxes to Address Climate Change

Many climate policy experts advocate for taxes on GHG emissions as the most economically efficient way to achieve needed emissions reductions.[30] Such taxes are often referred to as "carbon taxes," and like their close cousin of cap and trade, they have the effect of putting a price on carbon emissions, raising the cost of those emissions, and encouraging emitters to find ways to emit less. Implementation issues are similar to those for cap and trade, but there is no need for a market mechanism to facilitate trading. Setting the proper rate is the biggest challenge for a carbon tax, due to the uncertainty about the actual levels of emissions reductions that will be achieved for a given tax rate and the political and economic ramifications of setting the tax rate too high. In addition, a tax must consider the use of the revenues generated by the tax. A tax may be revenue neutral if proceeds are used to reduce other taxes. Alternatively, revenues from a carbon tax might be used to subsidize clean energy research or implementation.

---

[30] *See, e.g.,* Mann, *supra* note 25 at 10120; Richard D. Morgenstern, Reducing Carbon Emissions and Limiting Costs, Resources for the Future 166 (2002); Paul Krugman, *Building a Green Economy*, The New York Times Magazine, Apr. 11, 2010, at MM34.

Fifteen countries have imposed some form of carbon taxation. Some of these programs date back to the early 1990s and contain exceptions and complications that limit their effectiveness in lowering GHG emissions.[31] In Canada, the province of British Columbia adopted a simple fossil fuels carbon tax in 2008.[32] The level of the tax has risen to CA$38 per ton as of 2019. This carbon tax has helped achieve estimated emissions reductions of between 5% and 15%. In the US, carbon tax legislation has been introduced, occasionally with bipartisan sponsorship, but has not proceeded to a vote. The Energy Innovation and Carbon Dividend Act[33] would assess a payment for each ton of GHG emissions. In order to avoid anti-tax politics, the payment is called a "fee" rather than a tax. In order to address concerns about revenue neutrality and the disproportionate impact of a carbon tax on low-income energy consumers, this proposed legislation would return all revenues generated equally to all households in the US. The fee would start at $15 per ton of GHG emissions and increase by $10–15 annually until the emissions reductions goals of the statute were met. The emissions reductions targets would achieve roughly a 50% reduction in GHG emissions by 2035, as compared to 2016, and nearly a 90% reduction by 2050.[34]

F   Subsidies

A subsidy operates as a negative tax, which provides a commensurate economic inventive to engage in the activity that receives the subsidy. Indeed, many US subsidies take the form of so-called "tax subsidies," that is, special tax deductions and credits available to individual and business taxpayers that engage in the favored activity. By making lower-emitting activities more profitable, subsidies give a competitive advantage to enterprises with lower emissions, providing an economic incentive to adopt the lower-emitting technologies. Subsidies tend to be more popular politically than pollution taxes or fees. In addition to tax preferences, subsidies may take the form of direct grants and low-interest loans.

Subsidies may be considered economically inefficient, as they interfere with market allocations. In theory, a subsidy that simply offsets the externalities of the non-subsidized activity should be equally efficient as carbon tax based

---

[31]   *See* Shi-Ling Hsu, *Carbon Pricing, in* Legal Pathways to Deep Decarbonization in the United States 75 (M. Gerrard & J. Dernbach eds., Environmental Law Institute 2019).

[32]   *Id.*

[33]   Energy Innovation and Carbon Dividend Act of 2019, H.R. 763, 116th Cong. (1st Sess. 2019).

[34]   *See id.* § 9903.

on those externalities, but in practice subsidies are limited to specifically designated activities rather than all low-emitting activities. In practice, then, subsidies require a policy choice about which technologies will receive the subsidy, and which technologies will not. For example, renewable energy subsidies may or may not include subsidies for hydroelectric power or nuclear energy, even though those electricity-generation sources generally have greatly reduced emissions compared to fossil fuel generators. Energy subsidies may have the effect of lowering energy prices and increasing energy consumption, which may run counter to energy-efficiency goals.[35] Administration of grant and loan programs involves the granting agency in selecting specific firms and projects that will be the recipient of the government largesse.

Economists see the greatest benefit of subsidies in the case of a new technology seeking to displace an entrenched industry. A new technology may not be able to compete with an entrenched, "locked-in" industry even if the new technology is potentially cheaper and cleaner, due to the costs of switching and the economies of scale enjoyed by the existing industry. In this case temporary subsidies enable the new technology to get a market foothold sufficient to displace the entrenched industry.

Subsidies are already routinely used to support lower-polluting technologies in the United States, including technologies aimed at reducing GHG emissions. These existing subsidies include a federal investment tax credit for solar and wind projects that was adopted in 2006 and is scheduled to expire in 2022. Another example of GHG reduction subsidy is the federal tax credit for electric vehicles. State-instituted renewable portfolio standards, which require state-regulated electric utilities to source a specified percentage of their energy from renewable energy sources, may also be considered a form of renewable energy subsidy, though they might also be considered a technology mandate. These programs have been remarkably effective in reducing the GHG emissions from electricity generation and are discussed in greater detail in Chapter 3 of this book.

IV   GREENHOUSE GAS REGULATION UNDER THE EXISTING AUTHORITY OF THE CLEAN AIR ACT

While a comprehensive US regulatory response to global warming will likely require specific legislation, the existing provisions of the CAA include authority to regulate GHG emissions, and the EPA has exercised this authority. This

---

[35]   *See* Jim Rossi, *Electricity Charges, Mandates, and Subsidies, in* Legal Pathways to Deep Decarbonization, *supra* note 2, at 605.

section will provide a brief introduction to the structure of the CAA, a history of the EPA's partial regulation of GHG emissions, and a discussion of additional CAA authorities the EPA could use to expand its regulation of GHG emissions without new legislation.[36]

A   Overview of the Clean Air Act's Regulatory Provisions

The CAA authorizes the EPA to regulate "air pollutants." The Act incorporates a combination of direct emissions regulations, state implementation, technology-based emissions control requirements, emissions trading, and health-based air quality standards in an exceedingly complex regulatory scheme. The CAA includes several overlapping and duplicative systems of regulation. This brief introduction will not attempt to cover them all.

The basic structure of the CAA divides air pollution sources into two types: stationary sources and mobile sources. Stationary sources are regulated under Title I of the Act, which includes both the health-based NAAQS and technology-based emissions controls that apply in particular situations. Title II of the Act regulates mobile sources, which include motor vehicles as well as other transportation sources such as rail locomotives and aircraft. Title IV establishes a cap-and-trade program for reducing emissions of acid rain precursors. Title V establishes a permitting requirement for major sources, to be administered primarily by the states. Major sources subject to Title V permitting are those that emit more than 100 tons per year of "any air pollutant."[37]

The Title I NAAQS program was described above as an example of an environmental quality-based system of emissions limitations. The EPA designates a list of those air pollutants to be controlled, known as criteria pollutants. The EPA then establishes maximum concentrations in ambient air for these criteria pollutants. The primary NAAQS are set at a level to protect public health, while the secondary NAAQS are set at a level necessary to protect public welfare. Once the EPA establishes NAAQS, states develop plans (with EPA oversight) to regulate emissions within their state in order to achieve and maintain compliance with NAAQS within their state. Although Title I primarily regulates stationary sources, these implementation plans may include transportation control measures to reduce vehicle emissions.

Title I also includes a bewildering array of technology-based standards that may apply in particular circumstances. New source performance standards

---

[36]   This discussion is highly simplified. For a more in-depth introduction to the Clean Air Act, *see* The Clean Air Act Handbook, ABA Section on Energy, Environment, and Resources (J. Domkie & A. Zacaroli eds., American Bar Association 4th ed. 2016).

[37]   Clean Air Act (C.A.A.) §§ 501(2), 302(j), 42 U.S.C. §§ 7661(2), 7602(j).

apply to emissions of regulated pollutants from new and substantially modified facilities in industries designated by the EPA. In addition, new or modified major emitting facilities must also undergo new source review, conducted by the state as part of its SIP. New source review is intended to ensure that the source does not contribute to NAAQS violations in an area meeting the NAAQS standards or to ensure that the source completely mitigates its emissions if it is located in an area that violates standards. "Major emitting facilities" subject to new source permitting review are generally those that emit more than 250 tons of a criteria pollutant per year, or 100 tons per year for electric generating units.[38] Hazardous air pollutants may be subject to their own set of standards. Existing sources may also be subject to technology-based standards based on the need to achieve NAAQS in a non-attainment area, or based on EPA guidelines. Table 2.1 details the different technology-based standards that may apply under the CAA, roughly in increasing order of stringency.

Table 2.1

CAA § AND STANDARD	ABBREVIATION	APPLICABILITY	IMPLEMENTED BY
§ 172 Reasonably Available Control Technology	RACT	Existing sources in non-attainment areas	States' SIPs, based on EPA guidance
§ 111(d) Best System of Emissions Reductions for Existing Sources	BSER	Existing sources in EPA designated industrial categories	States' SIPs, based on EPA standards
§ 111(b) New Source Performance Standards, or Best Demonstrated Technology	NSPS or BDT	New sources in EPA designated industrial categories	States, through permitting for new sources, based on EPA standards
§ 165(a) Best Available Control Technology	BACT	Major new sources in areas attaining NAAQS subject to new source prevention of significant deterioration review, applies to all pollutants subject to regulation under CAA	States, through permitting for new sources, based on EPA guidance

---

[38]   C.A.A. § 169(1), 42 U.S.C. § 7469(1) (note that the "major" facility threshold for new source review is different than the definition that applies for Title V permitting—all facilities are subject to a 100 ton per year for Title V permitting).

CAA § AND STANDARD	ABBREVIATION	APPLICABILITY	IMPLEMENTED BY
§ 173(a) Lowest Achievable Emissions Rate	LAER	Major new sources in areas not meeting NAAQS subject to new source nonattainment review	States, through permitting for new sources, based on EPA guidance
§ 112 Maximum Achievable Control Technology	MACT	Hazardous pollutants designated by the EPA emitted by industries designated by the EPA	States, through Title V permits, based on EPA standards

Title II of the CAA addresses standards for mobile sources, which include automobiles and trucks as well as other mobile sources such as railroad locomotives, ships, and aircraft. The EPA may set pollutant emissions standards for engines used in motor vehicles under CAA § 202, and the EPA may also regulate motor vehicle fuels and fuel additives under CAA § 211. In general, the CAA preempts states from adopting motor vehicle emissions standards that differ from the federal standards; however, the CAA allowed the EPA to grant a preemption waiver to California, due to California's extreme air quality violations. When this waiver is in effect, California can impose stricter vehicle emissions standards than the EPA has adopted. Any other state may choose to adopt the stricter California standard to apply to vehicles sold in that state.

B   History of GHG Regulation Under the Clean Air Act

This section provides an overview of the history of the EPA's regulation of GHG emissions under the CAA, and a summary of the EPA's most significant GHG regulations.[39] As described above, the CAA gives the EPA a broad range of regulatory tools to address emissions of air pollutants, and directs the states to implement EPA air quality standards for criteria pollutants and emissions standards for regulated industry categories. The key finding that triggers regulation of a particular air pollutant is an EPA determination that a pollutant,

---

[39] This summary does not purport to be exhaustive. Additional EPA regulation of GHGs not covered in this section includes renewable fuels standards, greenhouse gas inventory reporting regulations, methane emissions regulations for oil and gas production, among other areas. For a more complete analysis of EPA GHG regulation, *see* Kyle Danish & Avi Zevin, *The Clean Air Act and Global Climate Change, in* The Clean Air Act Handbook 563–614 (J. Domkie & A. Zacaroli eds., American Bar Association 4th ed. 2016).

or industry emissions, "cause[s] or contribute[s] to, air pollution which may reasonably be anticipated to endanger public health or welfare."[40]

Despite the official recognition by the United States in the 1992 United Nations Framework Convention on Climate Change that global warming poses a threat to human health and welfare, the EPA made no attempt to regulate GHG emissions until it was forced to address the issue by the Supreme Court in *Massachusetts v. EPA*.[41] Since then, the scope and ambition of EPA regulation of GHG emissions has varied with the political winds of succeeding presidential administrations, leaving the current status of CAA regulation of GHGs very much in flux. What follows is a brief history of the EPA's regulatory actions under the CAA.

The CAA defines "air pollutant" to mean "any air pollution agent or combination of such agents, including any physical, chemical, biological, radioactive … substance or matter which is emitted into or otherwise enters the ambient air."[42] Section 202 of the CAA directs the EPA to adopt emissions standards for air pollutant emissions from motor vehicles that "contribute to air pollution which may reasonably be anticipated to endanger public health or welfare."[43] Even in 1970, Congress included effects on "climate" in the definition of "public welfare" endangerment.[44]

Given the statutory language supporting GHG regulation, it was only a matter of time before environmental interests sought EPA regulation under the CAA. A group of environmental NGOs petitioned the EPA to regulate GHGs under CAA § 202 in 1999.[45] In 2003, during the George W. Bush administration, the EPA denied the petition. The EPA reasoned, inter alia, that Congress could not have intended to include ubiquitous GHGs in the definition of pollutants subject to comprehensive regulation under the CAA. Eventually, the issue reached the Supreme Court, which held in *Massachusetts v. EPA* that the plain language of the CAA necessarily encompassed regulation of GHGs as air pollutants.

Although *Massachusetts v. EPA* settled the legal question of whether GHGs were air pollutants potentially subject to CAA regulation, it did not settle the factual question of whether the emissions of GHGs may reasonably be anticipated to endanger public health or welfare. The EPA avoided making a finding either way on this issue for the remainder of the Bush administration. When

---

[40]   *See, e.g.*, C.A.A. §§ 108(a)(1)(A), 111(b)(1)(A), 202(a)(1), 211(c), 42 U.S.C. §§ 7408(a)(1)(A), 7411(b)(1)(A), 7521(a)(1), 7544(c).

[41]   Massachusetts v. E.P.A., 549 U.S. 497, 508–509 (2007).

[42]   C.A.A. § 302(g), 42 U.S.C. § 7602(g).

[43]   C.A.A. § 202(a)(1), 42 U.S.C. § 7521(a)(1).

[44]   C.A.A. § 302(h), 42 U.S.C. § 7602(h).

[45]   *Massachusetts*, 549 U.S. at 510.

the new EPA administrator took office with the Obama administration in 2009, the EPA took up the question of whether GHGs caused an endangerment. On December 7, 2009, the EPA issued its endangerment finding, concluding that GHG emissions posed an endangerment to both public health and public welfare.[46] The court challenge to this finding was ultimately unsuccessful.[47]

The EPA Endangerment Finding triggered several regulatory actions on the part of the EPA. First, it obliged the EPA to adopt motor vehicle emissions standards under CAA § 202. Then, once GHGs became pollutants that are "subject to regulation under this Chapter," this triggered the BACT requirement for new source PSD review.[48] In addition, the EPA would be required to issue NSPS standards for any industry category that causes or contributes to the endangerment.[49] Once an industry category is so designated, CAA § 111(d) obligates the EPA to develop BSER guidelines for existing sources within that category as long as GHGs are not designated as a criteria pollutant. In addition to these emissions guidelines, the EPA was confronted with the question of whether the classification of GHGs as "air pollutants" subject to regulation under the CAA meant that new source review requirements would be triggered for any facility emitting more than 250 tons per year of GHGs. During the remainder of the Obama administration, the EPA addressed each of these issues by regulation. Each of these regulatory initiatives has subsequently been modified by the judicial system or has been subject to reconsideration during the Trump administration.

A summary of the EPA's most significant GHG regulatory action and current status of each of these regulatory initiatives follows. This is not an exhaustive treatment of every action taken by the EPA with respect to GHG emissions.

1   Motor vehicle emissions standards

Based on the endangerment finding, the EPA was obligated to issue motor vehicle GHG emissions standards under CAA § 202. Because the most significant GHG emissions from motor vehicles consist of CO_2, and because CO_2 emissions are entirely dependent on the quantity of hydrocarbon fuels burned, regulation of GHG emissions would in effect be a regulation of minimum

---

[46]   Endangerment and Cause or Contribute Findings for Greenhouse Gases Under Section 202(a) of the Clean Air Act, 74 Fed. Reg. 66,496, 66,496–546 (Dec. 15, 2009) (to be codified at 40 C.F.R. ch. I).

[47]   Coalition for Responsible Regulation v. E.P.A., 684 F.3d 102 (DC Cir. 2012), (rev'd in part on other grounds sub nom); Utility Air Regulatory Group v. E.P.A., 573 U.S. 302 (2014).

[48]   C.A.A. § 165(a)(4), 42 U.S.C. § 7475(a)(4).

[49]   C.A.A. § 111(b), 42 U.S.C. § 7411(b).

fuel economy standards. Under the Energy I and Security Act, the National Highway Transportation Safety Administration (NHTSA) is already required to establish motor vehicle fuel economy standards. Even before the endangerment finding, the State of California had requested an EPA waiver to issue its own set of GHG emissions standards.

These potentially overlapping fuel economy standards posed compliance issues for the automobile industry. The potential conflict was ultimately resolved in a negotiated rulemaking in which the EPA granted California's waiver, and the EPA and NHTSA issued identical fuel economy standards that were consistent with California's standards, resolving the potential regulatory conflicts. The phase one standards, for model years 2012 to 2016, were issued in 2010, and provided for average fuel economy standards for cars and light trucks equivalent to 34.5 miles per gallon for the 2016 model year.[50] The phase two standards, for model years 2017–2025, were issued in 2012, and provided for the equivalent of 54.5 miles per gallon in model year 2025.[51] The 2012 phase two standards provided for a midterm re-evaluation for model years 2022–2025. In April 2020, the Trump administration invoked the midterm review provision for phase two standards in order to relax emissions standards for model years 2022–2025.[52] These new standards are projected to result in average fuel efficiency for model year 2030 vehicles of 40.5 miles per gallon. Upwards revision of the mileage standards appears to be likely in the new Biden administration.

2   Emissions standards for new sources

As outlined above in the overview of the CAA, emissions sources, including power plants, are potentially subject to several overlapping permitting requirements and technology-based emissions standards. New major emitting sources are subject to new source review and permitting requirements, including NSPS and BACT, for pollutants regulated under the CAA. Once the EPA regulated GHG emissions from motor vehicles under CAA § 202, GHG emissions from new sources became subject to these emissions standards.

The initial regulatory issue the EPA had to confront was the question of whether GHG emissions were to be counted for the purposes of applying the 250 ton per year and 100 ton per year thresholds for determining whether

---

[50]   Light-Duty Vehicle Greenhouse Gas Emissions Standards and Corporate Average Fuel Economy Standards; Final Rule, 75 Fed. Reg. 25, 324 (May 7, 2010).

[51]   2017 and Later Model Year Light-Duty Vehicle Greenhouse Gas Emissions Standards and Corporate Average Fuel Economy Standards, 77 Fed. Reg. 62, 624 (Oct. 15, 2012).

[52]   The Safer Affordable Fuel-Efficient (SAFE) Vehicles Rule for Model Years 2021–2026 Passenger Cars and Light Trucks, 85 Fed. Reg. 24, 174 (Apr. 30, 2020).

a facility was a "major" source that triggered permitting and emissions standards. The EPA had traditionally interpreted the "any air pollutant" language of the major emitting facility definitions to refer to any air pollutant subject to regulation under the CAA, whether the pollutant was a designated criteria pollutant subject to the NAAQS or not.[53] This interpretation posed an administrative problem, as the quantities of GHGs produced by burning fossil fuels was so large that millions of small facilities, including schools and apartment buildings with oil heat, would be subject to permit procedures.[54] The EPA sought to avoid this problem by issuing the Tailoring Rule, which attempted to modify the statutory permitting triggers to 50,000 tons per year of CO_2e when fully phased in. The Supreme Court ultimately struck down the EPA's interpretation of the permit trigger thresholds to include non-criteria pollutants such as GHGs.[55] As a result, Title V permitting requirements and new source permitting would apply only to those facilities that would require these permits anyway for criteria pollutants—so-called "anyway sources." A facility would not be subject to permitting solely based on its emissions of GHGs.

However, most utility-scale electric generating plants are "anyway sources" which require new source review based on their emissions of criteria pollutants such as sulfur dioxide, nitrogen oxides, carbon monoxide, and particulates. While the challenge to the Tailoring Rule was pending, the EPA proceeded to establish New Source Performance Standards for two industries—the electric generating industry, and the oil and gas production industry. The NSPS for electric generation units were issued in 2015, and set emissions rates for GHGs that were based on the most efficient combustion technologies available, plus partial carbon capture and storage for new coal-fired units.[56] In late 2018, as part of the series of regulatory rollbacks proposed by the Trump administration, the EPA issued a proposed rule relaxing the NSPS for GHG emissions from electric generating units.[57] These revised standards, which were not finalized before the end of the Trump administration, would have relied solely on maximizing generation efficiency, and would remove any reliance on carbon capture and storage.

---

[53] *See* Prevention of Significant Deterioration and Title V Greenhouse Gas Tailoring Rule, 75 Fed. Reg. 31,513, 31,550 (June 3, 2010).

[54] *Id.* at 31, 555.

[55] Utility Air Regulatory Group v. EPA, 573 U.S. 302 (2014).

[56] Standards of Performance for Greenhouse Gas Emissions from New, Modified, and Reconstructed Stationary Sources: Electric Utility Generating Units, 80 Fed. Reg. 64,510 (Oct. 23, 2015).

[57] Review of Standards of Performance for Greenhouse Gas Emissions From New, Modified, and Reconstructed Stationary Sources: Electric Utility Generating Units, 83 Fed. Reg. 65424 (Dec. 20, 2018).

3   BSER regulation of existing electric generating units

Having established NSPS for the electricity utility generating unit industry category, the EPA was required under CAA § 111(d) to establish BSER standards for GHG emissions from existing sources in that category, to be applied and implemented by states. This regulation took the form of what the Obama administration called the "Clean Power Plan."[58] The Clean Power Plan took the form of a complex set of emissions reductions goals for each of 47 specified states. The Clean Power Plan targeted the achievement of a 32% reduction in GHG emissions from the US power sector by 2030, which was consistent with the nation's intended nationally determined contribution to GHG emissions reductions under the Paris Agreement. These emissions reductions goals went beyond unit-by-unit emissions reductions goals, and contemplated achievement of state-wide electricity generation-related carbon emissions through increased reliance on low-carbon emitting and renewable energy sources. The Clean Power Plan specifically relied on emissions trading, including interstate emissions trading, to achieve the reductions contemplated. The Clean Power Plan was never implemented, however, as the Supreme Court issued a stay of the regulation in 2016,[59] and the Trump administration repealed the Clean Power Plan and issued new BSER GHG standards based solely on efficiency improvements at existing power plants in 2019.[60] This repeal of the Clean Power Plan was vacated by the District of Columbia Circuit in January 2021,[61] and the section 111(d) Best System of Emissions Reductions regulations will be reconsidered by the new Biden administration EPA.

C   Potential Further Mitigation Regulations Under Existing CAA Authority

Although regulatory rollbacks under the Trump administration have not resulted in the EPA disclaiming authority to regulate GHGs at all, the repeal of the Clean Power Plan and relaxation of the stringency of new source requirements for GHGs have resulted in a set of GHG regulations that are unlikely to significantly mitigate US GHG emissions. The incoming Biden administration, which has declared action on climate as one of its highest priorities, will likely reinvigorate and expand on the regulatory authority, invoked during the Obama administration, to establish mobile source standards and technology-based stationary source emissions standards. Emissions standards

---

[58]   Carbon Pollution Emission Guidelines for Existing Stationary Sources: Electric Utility Generating Units, 80 Fed. Reg. 64,662 (Oct. 23, 2015) ("Clean Power Plan").
[59]   West Virginia v. E.P.A., 136 S.Ct. 1000 (2016) (mem., order granting stay).
[60]   *See* Repeal of Clean Power Plan, *supra* note 11, at 32522.
[61]   American Lung Ass'n v. EPA, 985 F.3d 914 (D.C. Cir. 2021).

under section 111(b) and (d) could be expanded to other industrial sectors that are significant sources of GHG emissions, such as cement production. In addition to these authorities, several other sections of the CAA potentially provide the EPA with additional authority to regulate GHG emissions, potentially more comprehensively than authorities invoked to date. These authorities include the NAAQS program, the authority to regulate fuels under CAA § 211, and the authority to address cross-boundary international air pollution under CAA § 115.

1   GHGs as a NAAQS criteria pollutant

Under CAA § 108, the EPA can expand the list of criteria pollutants subject to regulation under the NAAQS program. This section directs the administrator of the EPA to designate for the NAAQS program those pollutants:

(A)   emissions of which, in his judgment, cause or contribute to air pollution which may reasonably be anticipated to endanger public health or welfare;

(B)   the presence of which in the ambient air results from numerous or diverse mobile or stationary sources; and

(C)   for which air quality criteria had not been issued before December 31, 1970 but for which he plans to issue air quality criteria under this section.

The EPA thus has authority to designate GHGs as criteria pollutants based on the 2009 endangerment finding that formed the basis of motor vehicle emissions standards.

Listing GHGs as criteria pollutants subject to the NAAQS program would trigger several administrative and implementation actions. First, the EPA would be charged with developing ambient air quality standards for GHGs sufficient to protect the public health and welfare. Next, each state would be required to amend its SIP to demonstrate achievement of the concentrations established by the EPA within three years. It would be unlikely that any state could demonstrate compliance with this requirement, as GHG concentrations in the atmosphere are well dispersed through the global atmosphere, and concentrations within one state will not be substantially affected by emissions reductions within the state. The failure of states to demonstrate compliance with a GHG NAAQS could empower the EPA to adopt a federal implementation plan imposing limits on stationary sources throughout the nation as necessary to achieve compliance with the NAAQS.

A listing of GHGs as criteria pollutants is clearly authorized by the language of the CAA. However, several factors make the NAAQS program a potentially unwieldy tool for addressing GHG emissions. The resulting shift of national implementation power from states to the EPA runs counter to the Act's environmental federalism structure. The EPA would also face the same challenge as the states in developing a national FIP that itself was capable of achieving a health and welfare-based concentration of GHGs. In addition, once the EPA has rejected a state's SIP and imposed a FIP, it must sanction the state by denying federal highway funding to the state (with some exceptions for mass transit and carpool lane funding).[62] An all-state FIP would essentially end federal highway funding, unless Congress acted to restore it.

2   International air pollution controls under CAA § 115

Section 115 of the CAA authorizes the administrator to require a state to revise its SIP to address air pollutant emissions that result in an endangerment to the public health or welfare in a foreign country.[63] There are two conditions on the exercise of this authority: 1) the determination of endangerment must be based on a report of a duly constituted international agency, or be in response to a request made by the Secretary of State, and 2) the foreign nation must be under a reciprocal obligation to take equivalent mitigation measures. The IPCC assessment reports establishing the hazards of global warming as a result of uncontrolled emissions of GHGs likely satisfies the first requirement. The second condition will require reference to an international agreement providing for commensurate emissions reductions in other nations. The EPA might rely on the 1992 United Nations Framework Convention on Climate Change, or the 2015 Paris Agreement to establish such reciprocity. The EPA could require states to amend their SIPs to show achievement of the appropriate level of GHG emissions reductions, and presumably could impose a FIP to implement reductions within states that fail to comply.

3   Regulation of mobile source fuels under CAA § 211

Section 211 of the CAA authorizes the EPA to "control or prohibit the man-ufacture ... or sale ... for use in a motor vehicle" if the EPA determines that "any emission product of such fuel ... causes, or contributes to air pollution ... that may reasonably be anticipated to endanger the public health or welfare."[64] As the EPA has already made this endangerment finding specifically in order to regulate emissions from motor vehicles, the scientific predicate for invoking section 211(c) has already been satisfied. Under the language of this section, the EPA might adopt a wide range of controls on the sale and use of fossil fuels for transportation purposes. These controls might range from "controls" such as caps on quantities of gasoline, diesel, or jet fuel, to an outright prohibition against the use of fossil fuels for transportation purposes. In order to invoke this authority, the EPA must consider medical, scientific, and economic evi-

---

[62]   Clean Air Act (C.A.A.) § 179(a), (b)(1), 42 U.S.C. § 7509(a), (b)(1).
[63]   C.A.A. § 115, 42 U.S.C. § 7415.
[64]   C.A.A. § 211(c)(1), 42 U.S.C. § 7545(c)(1).

dence concerning alternative means of protecting public health and welfare, and the EPA must also find that the proposed control on fossil fuels will not result in switching to other fuels that will cause an equal or greater endangerment to the public health and welfare.

As transportation fuels account for 28% of US GHG emissions, section 211(c) provides the EPA with a potentially powerful mitigation tool.

CONCLUSION

As anthropogenic climate change is caused by emissions of contaminants into the ambient environment by human activities, many of the traditional policy tools for controlling harmful pollution discharges may be adapted to mitigate GHG emissions. These tools include direct regulation of pollution rates and quantities, as well as economic and market-based incentives to reduce high-emitting activities and substitute lower-emitting activities. The US CAA provides existing authority for GHG regulation, which has been exercised incrementally by the EPA since the Supreme Court declared that GHGs were air pollutants subject to regulation under the Act in 2007. Additional regulation of emissions under the CAA remains possible.

3. Introduction to energy law

Karl R. Rábago and Radina Valova

I INTRODUCTION

Energy law is a critical subset of climate law. The majority of global greenhouse gas emissions come from the energy sector, which broadly includes three sub-sectors: electricity production, transportation, and fossil fuels used in buildings and industrial processes.[1]

Energy sector emissions come not just from the combustion of fossil fuels and biomethane, but from the entire energy supply chain and lifecycle of each component, from extraction to transportation, distribution, and waste disposal (e.g., management of coal ash). The fossil fuel extraction process, for example, releases methane during coal and gas development, and carbon dioxide when gas is flared to reduce pressure in drilling wells, as well as methane due to leakage in gas transmission and distribution pipelines. Given the energy sector's impacts on climate change, to the extent energy *law* impacts energy production, delivery, and use, it *is* climate law. From a policy perspective, therefore, electricity regulation, and increasingly gas and petroleum regulation, have been common vehicles for advancing climate regulation and policy goals.

So, over the past several decades, climate has become an essential consideration in energy law. This is remarkable, as in other fields of law, because the impacts of climate change, unlike most environmental impacts, are indirect, cumulative, delayed, and long-lived. The extent to which energy law has been reshaped around the climate issue is perhaps most profound because of the overwhelming human dependence on fossil fuels, and the magnitude of transformative change in energy that climate change compels.

---

[1]   Industrial process uses of energy include direct energy combustion—mostly gas—for thermal energy and self-production of electricity and recovery of waste heat, as well as farming, which makes direct use of energy, indirect use through fertilizer and other inputs, and impacts climate through soil and land use changes. This section does not further address fossil fuels used in industrial processes, but does address the combustion of gas and fuel oils for heating and other services in buildings.

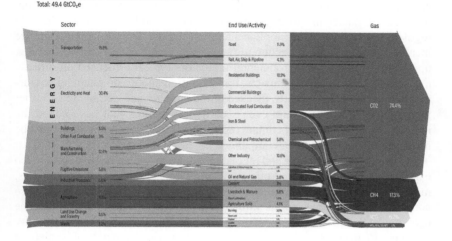

Figure 3.1   World greenhouse gas emissions in 2016

World Greenhouse Gas Emissions in 2016, WORLD RESOURCES INSTITUTE (Feb. 3, 2020). Source: https://files.wri.org/d8/s3fs-public/world-greenhouse-gas-emissions-sankey-chart-2019_0.pdf.

Climate impacts are arguably today's most important single metric for energy law, revealing a fundamental tension between the benefits that energy provides and the consequences of fossil energy use. The law, which seeks to account for private and public property rights, economic efficiency, the public interest, and just distribution of benefits and burdens, has been increasingly relied upon as the mechanism for addressing that fundamental tension.

What happens, for example, when restrictions on carbon emissions limit or end the operation of coal-fired power plants, which in turn represent billions of dollars in investment and have decades of potential remaining useful life? And how does the law weigh the impacts of lost jobs in coal mining towns against the suffering of flood victims in coastal areas? How do the interests of future generations get weighed against those of people today? Who has standing to seek limits or prohibition based on not just the potential nuisance impacts of energy development and industry operations, but also based on the consequences of incremental contributions to global climate change?

While many laws in the United States and other nations require regulatory consideration of climate impacts and carbon emissions in energy regulation, the task of balancing such competing interests adds new dimensions to regulatory oversight. The implementers of climate law and regulation are increasingly public utility economic regulators that oversee utility planning, operations, and rates; environmental regulators typically focused on directly measurable emissions and impacts; transportation and urban planning authorities that in the past typically focused on geographic and demographic rather than atmospheric issues; and even securities regulators that have added the risks of climate change and climate regulation to the list of issues that must be disclosed by publicly traded companies.

In many ways, legislative and regulatory attention to climate matters has also expanded awareness of how climate change impacts other policy concerns, making a strong case for a broader definition and conceptualization of these concerns. For example, public utility rates are broadly designed to advance economic efficiency. But ratemaking that ignores the so-called externality impacts of carbon emissions—long-term and unpriced impacts on society at large—will distort production and consumption decisions, causing more of the product to be made or used than is efficient. As another example, the production of energy without carbon emissions, such as by wind and solar energy technologies, has led to recognition of property rights in the *non-creation* of carbon emissions when such technologies operate. Today, renewable energy certificates and carbon credits—often denominated in or converted to tons of carbon emissions *avoided*—are widely traded commodities around the world.

The sections below cover three essential legal touch points along the energy system: the basics of energy regulation—public service and administrative law; and laws governing the development of energy projects, the regulation of energy in the buildings sector, and laws governing the transportation sectors, all of which are intended to move the current inefficient and polluting energy system to a cleaner, more affordable, and resilient future. Across each of these regulatory touch points, just transition from a fossil-intensive to a more renewable and sustainable energy system is a key element of the clean energy and climate transition: Some jurisdictions are developing policies to ensure that communities that have historically been disproportionately impacted by the environmental and health impacts of fossil fuels benefit from the move toward clean and renewable energy. Equitable access to climate-friendly, clean, and sustainable energy technologies and services is becoming a major issue of policy, legislation, and regulatory law.

II   THE BASICS OF ENERGY REGULATION

Energy regulation encompasses the full lifecycle of energy use, from the extraction or production of raw materials (e.g., extracting fossil fuels, producing biomethane from waste), to transport and delivery (e.g., pipeline leakage and safety, conversion losses, transmission system conductive losses), to end-use by consumers (e.g., building energy codes, utility rates and tariffs), or the power generation industry (e.g., utility resource planning, energy market design and operation). The energy lifecycle includes multiple climate impacts, including gas flares during the extraction process, methane leaks along gas pipelines, the combustion of fossil fuels for various energy services, and the energy it takes to produce and deliver energy.

Various agencies are responsible for regulating each of these steps and processes. State or federal environmental agencies typically govern the byproducts of the energy lifecycle, from air and water emissions to coal ash and nuclear waste disposal. Public utility/service commissions (PUCs or PSCs) govern utilities' planning, resource acquisition decisions, and the ultimate production and delivery of energy (electric or gas) to end-use customers, and serve as the primary energy regulators, at least when measured by the dollar.

Because this section focuses on the operation and regulation of energy use in buildings and the electric and gas utilities that provide these services, it primarily describes the regulatory role of PUCs/PSCs. Section III below describes the various regulations applicable to the development and environmental governance of energy.

A   Administrative Law and Public Participation in Energy Regulation

Most energy regulation happens at the administrative agency level. Administrative agencies are empowered by national or state legislatures to perform specific duties and achieve specific mandates. Agencies promulgate their own regulations in furtherance of their legislative mandates, including specific procedural rules that generally must comport with state or national administrative procedure laws.

The regulatory process typically includes the following steps: The agency issues draft decisions, rules, or policy proposals, and then solicits public comments and reviews those comments before issuing final rulings. In order to be validly enforceable, an agency's ruling must be based on the administrative record developed throughout this process and must comply with the agency's procedural requirements (e.g., adequate public notice regarding the rulemaking; sufficient opportunity for public participation).

Although the administrative process for rulemaking is public, some agencies—including PUCs—set participant "standing" requirements, which place limits on who is permitted to participate in the regulatory process. This is a key point at which the jurisdiction's recognition of climate issues comes into play. Usually, this debate arises in disputes over whether parties seeking to have climate impacts addressed have raised issues within the jurisdictional authority of the administrative agency, or germane to a jurisdictional issue and, therefore, whether such parties may properly challenge agency decisions. The ultimate question is whether climate issues meet the requirements relating to standing and justiciable interest established by state and federal courts. As such, the process and results are not universally inclusive.

The energy law administrative process also requires a deep understanding of the procedural rules and substantive issues inherent in energy regulation, and stakeholders must have the time, resources, skills, and substantive competence to participate in often arcane and drawn-out proceedings over a period of months or even years. Significant issues arise, for example, over the appropriate depreciation schedules for fossil-fired assets, the appropriate discount rates to be applied to long-lived assets, and the relative decisional weight to be accorded to bill versus rate impacts.

As a result, participation tends to be limited to stakeholders who have the expertise and resources, which often leaves out representation from marginalized and frontline communities—particularly those who have historically been disproportionately impacted by the fossil fuel economy. Some jurisdictions are making an effort to address this challenge, such as by providing intervenor funds to support stakeholder participation in typically rarified regulatory proceedings.[2]

B   Public Service Law and the So-called Regulatory Compact

The administrative process described applies to each of the steps in the energy lifecycle, including the regulations governing oil and gas drilling companies, electricity generation facility operations, electric and gas transmission owners, and distribution utilities' operations. However, *utility* law differs substantially from other types of regulatory governance due to one central element: Energy utilities operate as regulated monopolies.

The extent to which utilities operate as monopolies varies from jurisdiction to jurisdiction. In some places, featuring "vertically integrated monopolies," the utility owns facilities and performs functions associated with generation

---

[2]   *See, e.g.*, the California Public Utilities Commission's Intervenor Compensation Program, https://www.cpuc.ca.gov/icomp/.

or production, transmission, distribution, and customer service. In others, the monopoly franchise has been limited to only transmission and distribution functions (e.g., in Texas) or to distribution functions (e.g., in New York).

Depending on the jurisdiction's specific market structure, utilities may be responsible for generating and/or purchasing, transporting, and delivering electricity and gas[3] to consumers, and for managing the distribution grids they own and operate. In some jurisdictions, utilities own and operate the system from the point of producing electricity (power generation), through long-distance transmission infrastructure, all the way to delivery to the end-use customer through the distribution grid—these are "vertically integrated" energy markets. In other jurisdictions, utilities are permitted to own and operate only the distribution grid and cannot own power plants or transmission lines—"deregulated" energy markets.[4] In either case, PUCs have the responsibility of regulating utilities in the public interest: They must ensure that utilities provide safe and reliable service at just and reasonable rates for electric and gas consumers.

In regulated markets, regulators typically exercise oversight on planning and infrastructure investment decisions (such as constructing new power plants), while in deregulated markets, such decisions are left to private investors. Interstate markets for gas and electricity are regulated by the federal government, typically by the Federal Energy Regulatory Commission (FERC) in the United States. At the state level, energy market regulators have increasingly called upon the utilities they regulate to account for the climate impacts of investment and operational decisions, to operate in ways that reduce or minimize greenhouse gas emissions, to prepare for the potential impacts of climate change, and to otherwise account for the climate-related policy priorities of the respective jurisdiction. Federal energy regulators have been slower to respond and, in some cases, have actually limited state action on climate issues over concerns about impacts on interstate commerce.[5]

When utilities operate as monopolies in their service territories, they enjoy an exclusive right to provide service to customers within that territory, primar-

---

[3] Some utilities also offer steam service, which provides heating or hot water, and has some specialized applications, such as in hospitals.

[4] States in which utilities are not permitted to own generation and transmission are commonly termed "deregulated" states, though regulation remains for many aspects of the businesses. Approximately half the states in the US are deregulated, in response to a financial crisis in the energy sector in the 1970s and 1980s. Deregulation is intended to create more competition in the energy production and transmission sectors, in furtherance of increased economic and energy efficiency and more affordable rates for consumers.

[5] Hughes v. Talen Energy Mktg., 136 S. Ct. 1288 (2016) (decision relying on a blended jurisdiction framework).

ily because when the electric utility industry was founded, the economics of installing and operating electric service infrastructure to provide electric or gas service favored a regulated monopoly structure over unrestrained competition. Under such economic circumstances, commentators and courts have spoken of the "natural monopoly." The thesis of the "natural monopoly" concept is that under certain circumstances, the provision of certain goods or services will be most economically efficient and therefore societally beneficial when performed by a business or organization with the exclusive right to serve all customers within a defined geographic area. Shortly after the advent of electricity generation, integrated utility companies successfully argued that it would be uneconomical and inefficient to have multiple companies building duplicates or triplicates of power stations, transmission lines, and poles and wires within a single geographic area. The trade-off for monopoly power was the accession to comprehensive regulation, including price regulation. This trade-off has been dubbed the "regulatory compact" to justify the departure from aspirational free-market principles, though the relationship does not bear the hallmarks of a contractual arrangement founded on legally enforceable mutual obligations.

Under the traditional regulatory model, PUCs granted to utilities charters or certificates of public convenience and necessity and allowed them to operate as regulated monopolies within their franchise service territories. In many locations, the original certificated companies have been operating as a monopoly for a hundred years or more. Utilities are permitted to recover from their customers the prudent and "used and useful" costs of providing service, as well as a reasonable return on investment for their shareholders—return "*of*" and *on* investment." In exchange for the right to monopoly ownership and operation, utilities are required to meet two primary obligations: PUCs require that utilities offer service at "just and reasonable rates," and that they meet certain standards, such as that service is "safe and reliable" or "safe and adequate." In a world characterized by a rapidly changing climate, both of these obligations are increasingly being revisited and modified. For example, is it reasonable to expect customers to pay for generation plants that emit significant amounts of carbon dioxide as a result of fossil fuel combustion, or to pay more for energy from non-emitting sources? Are coal-burning power plants safe?

The regulation of investor-owned energy utilities[6] is not simply about setting prices, known as rates, for services. Rate-making involves a wide range of issues, from the companies' capital structure and overall corporate financial governance, to rate design, to how efficiently they operate, both in terms of

---

[6]   This chapter primarily focuses on investor-owned utilities, as compared to municipally- or cooperatively-owned utilities.

their day-to-day management and in regard to energy use. Some regulatory issues are more clearly connected to climate law than others, but ultimately, every aspect of regulating the utility sector impacts how and how much utilities invest in energy assets, which invariably has an impact on climate.

Several issues deserve particular attention. The question of how utility rates are designed is an important element of climate policy: rates send price signals that can either encourage or discourage customers to use more or less energy, or to adopt or refrain from adopting energy efficiency, solar, energy storage, and other clean energy technologies. For example, rates that decrease with the amount of energy used tend to encourage greater consumption, while per-unit rates that escalate with the usage level tend to discourage excessive consumption. Higher volumetric rates—rates based on usage level—encourage energy conservation by customers that can afford to make improvements. Higher per-customer charges (monthly charges that don't vary based on usage level) make investment in efficiency and distributed resources, like rooftop solar, less economical.[7] Without customer-level adoption of clean energy, it will be difficult for state and national governments to achieve ambitious climate targets, so the issue of what kinds of rate designs create the strongest incentives for clean energy without being unfair is a matter of considerable debate.

The majority of utilities in the US are investor-owned.[8] And for most of those utilities, profitability is determined by throughput—the more energy sold, the higher the profits for shareholders. As a result, there is an inherent conflict between the fiduciary obligations of utility managers and the interests of efficient use of energy and conservation. Where electricity is generated by combustion of fossil fuels, that means there is an inherent conflict between utility obligations to shareholders and the reduction of greenhouse gases. Today, much of regulatory practice is therefore focused on questions like: How should utilities or generators make electricity? What role can utilities play in advancing efficient use and the transition to clean energy resources? What changes are needed in the traditional model of utility profit generation?

Most regulated utilities are required to offer clean energy programs, such as incentives to encourage customers to adopt energy efficiency, renewables, and, of increasing recent interest, alternatives to gas heating and cooling, such as heat pumps. In the case of vertically integrated utilities, additional governance also includes integrated resource planning: ensuring that they

---

[7] For more information on rate design, *see Pricing and Rate Design*, REGULATORY ASSISTANCE PROJECT, https://www.raponline.org/key-issues/pricing-rate-design/ (last visited Apr. 18, 2021).

[8] *Investor-owned utilities served 72% of U.S. electricity customers in 2017*, U.S. ENERGY INFORMATION ADMINISTRATION (Aug. 15, 2019), https://www.eia.gov/todayinenergy/detail.php?id=40913.

are investing in the least-cost generating assets that achieve state energy and climate objectives. More advanced regulatory tools include performance-based ratemaking, which requires utilities to meet certain policy outcomes, such as greater penetration of renewables or reductions in greenhouse gas emissions, and rewards the utilities with financial incentives for meeting those goals (or punishes failures). In the wake of severe weather events, like hurricanes and superstorms, regulators and legislators have also called on utilities to plan for and invest in infrastructure improvements to ensure energy system resilience during and after the severe weather that is sure to come.

When humans lacked a widespread comprehension of the climate consequences of dependence on fossil energy to provide basic services such as heat, light, and support for commercial and industrial activities, the role of energy regulators was narrowly economic and financial. Utility investments and operations had to provide the best "bang for the buck" while ensuring safe, reliable, and universally available access to service. Today, the roles and responsibilities of economic regulators have greatly expanded as the climate externality has been internalized through legislation, policy, evolving markets, and the need to respond to climate crises. The pace and extent of this internalization process is the stuff of much utility regulatory law and practice today, and will likely continue to be so for the future.

III   LAWS GOVERNING GREENHOUSE GAS EMISSIONS FROM THE POWER GENERATION, BUILDING, AND TRANSPORTATION SECTORS

Economics, including energy economics, is a function of both supply and demand. Because most end-use energy is the product of fossil fuel combustion, the ways and efficiency with which humans use energy is a driver of carbon and other emissions. Gas and electric utilities, along with the petroleum industry, are the major sources of climate-changing energy supply to homes, businesses, and transportation end-uses. In each case, energy use is incidental to human demand for services—warmth, cooling, mobility, agriculture, processing, or manufacturing. This section introduces the issues of law and regulation impacting electric power generation, use of energy in buildings, and use of energy for transportation.

Regulations governing energy project development include requirements intended to protect against environmental harm from the construction of a power plant to its emissions and its ultimate decommissioning. A multitude of agencies, statutes, and regulations govern the lifecycle, including environmental quality review laws, clean air and water rules, and waste management regulations.[9] In addition, local municipal governments typically have their own set of requirements, in addition to those set by national or state laws and administrative agencies, relating to siting, operational impacts like noise and vehicle traffic, and other areas of local jurisdictions.

As a first matter, it is important to note that this description is partitioned for convenience of discussion. In the real world, all economic and energy sectors interact with each other, necessitating an appreciation of both interactive and systems impacts. For example, a shift from petroleum-based transportation to electric transportation almost entirely eliminates end-use emissions from the mode of transportation—an electric automobile has no exhaust pipe. But how the electricity for the automobile is made matters a great deal. Traps for the unwary abound. For example, a law providing tax breaks for electric vehicle purchases could incrementally increase fossil energy use and related emissions due to the increased demand for electricity if that electricity is generated from fossil fuel combustion.

A   The Regulation of Energy in the Power Generation Sector

Several technological options exist for electricity generation. These include combustion of fossil fuels—coal, methane gas, and oil; nuclear reactions; and renewable energy—wind, sun, hydropower, and biomass. They also include large-scale power generators and smaller, distributed generators (DG), such as rooftop solar, fuel cells, and small wind turbines. DG is part of a broader set of distributed energy resources (DER) that not only create energy, but also manage, store, and convert it. DER technologies include DG, distributed energy storage, fuel cells, energy efficiency and conservation, energy management and demand response, and even electric vehicles when they are plugged into the electric grid.

The decisions about which kinds of power generators to build and operate in order to meet customer demand for electric service is a function of economics and financing, siting challenges, technological performance and reliability, environmental standards, and regulation and law. In some states and regions ("vertically integrated"), a single utility builds, owns, and operates its generation resources to serve all the customers in its designated service territory, though it may be required to purchase energy from certain renewable energy and cogeneration facilities under US federal law. In large parts of the US

---

9   As discussed in Chapter 2, *Climate Law Primer: Mitigation Approaches*, one key focus in the development of federal climate change mitigation policy has been how to use authority under the CAA to regulate GHG emission from new and existing energy sources.

("deregulated"), the generation sector is more competitive, and electricity from utility and non-utility generators is delivered to grid operators for subsequent resale and delivery to distribution utilities.

The extent to which generation decisions have been vested in regulators or market structures informs how climate issues can be addressed in power generator development, operation, and retirement decisions. In most US states, legislatures have enacted Renewable and/or Clean Energy Portfolio Standards dictating that an increasing share of electricity must come from non-fossil sources such as renewables and nuclear energy. Ironically, legislatures in coal-rich states have also considered laws maintaining a share of coal generation in the statewide system mix due to the adverse economic impacts of plant closures.

Where electricity markets are highly deregulated, a clean energy standard is typically imposed on retail sellers of electricity and can be met through some combination of energy supply contracts and renewable energy certificates (RECs). RECs are tradable embodiments of the renewable energy attributes of the underlying generation—type of technology, avoided emissions, avoided water consumption, and date and time of generation—created when renewable energy is generated. RECs can be bought and sold independently from the underlying physical electricity they were created with.

Where utilities are vertically integrated, the common approach to implementing a clean energy standard is through planning and construction requirements imposed on utilities. And in some cases, hybrid approaches are used—allowing a utility to comply with a clean or renewable energy standard through a least-cost mix of energy efficiency, renewables construction, power purchases from non-utility generation, and purchases of RECs from qualified resources. Flexibility in compliance is increasingly important as leading states move toward economy-wide energy policies—targeting greenhouse gas emissions reductions by electric and gas utilities, the transportation sector, industry, and in buildings.

Planning oversight is especially powerful as a tool for reducing greenhouse gas emissions from the electricity sector because power plants are extremely long-lived assets (multiple decades) and may require considerable complementary investments in transmission and interconnection infrastructure. Regulators seldom substitute their planning decisions for those by the regulated entity, and instead impose requirements on how resource options are evaluated.

To evaluate long-term costs and benefits, future impacts are often levelized to the value of present-day dollars in order to allow comparison of disparate technologies. The discount rate used is a critical variable. A high discount rate reduces the impact of future costs, such as fuel costs, and of future benefits, such as avoided carbon emissions. Regulators increasingly call upon utility planners to evaluate the benefits of avoided carbon emissions and contributions to climate change with societal discount rates that are substantially lower than the private discount rates typically used by utilities. Lower discount rates are therefore an economic tool for putting a higher value on the future.

Regulators increasingly call on utility planners to address damage costs of climate change through use of a social cost of carbon adder applied to cost estimates. For high-carbon resources, like coal, a cost of carbon adder will render future coal plant additions an extremely unlikely component of a utility resource plan. Such adders impact the outcome of complex modeling exercises used to select the resource mix—the kinds and sizes of electricity generators—installed and used to satisfy demand for electricity.

A number of states are also considering the role of DER in the overall power generation mix. More and more customers are adopting rooftop solar, energy storage, heat pumps, and other means of obtaining on-site energy services. The traditional utility planning process does not take such customer generation into account—as a simplified matter, utilities traditionally forecast their customers' anticipated energy usage and either build or buy sufficient large-scale generation to meet that forecasted demand. The traditional utility planning process also does not take into account the electric grid upgrades that will be necessary to accommodate the rapid interconnection of large amounts of DER required in order to meet state energy and climate goals. More progressive planning processes are beginning to emerge, such as planning for distribution system upgrades to accommodate more DER on the grid.[10]

Finally, open and transparent consideration of climate risk in utility planning has the effect of putting the utility and its shareholders on notice about those risks. As a result, future law and regulation limiting carbon emissions cannot be said to come as a surprise when high-carbon resources are effectively priced out of operations. Not only does this impact the prudence of current investment decisions about new sources of generation, but also the reasonableness of continued reliance on carbon emitters built in years past.

B   The Regulation of Energy in the Buildings Sector

For most of us, buildings are the most significant way we interact with energy, and therefore, with the uses of energy that increase greenhouse gas emissions. Buildings use energy in several ways. Methane gas and fuel oil are directly used in boilers or furnaces to generate heat and provide heat for cooking.

---

[10]   *See, e.g.*, Zachary Gerson, *Department of Public Utilities Proposes Rethink for Distribution System Planning and Interconnection Costs*, ENERGY AND CLEAN TECH COUNCIL (Oct. 30, 2020), https://www.energycleantechcounsel.com/2020/10/30/department-of-public-utilities-proposes-rethink-for-distribution-system-planning-and-interconnection-costs/.

Almost all the rest of building energy use is electricity, for lights, refrigeration, appliances, and "heating, ventilation, and air conditioning" (commonly called HVAC). As a result, how well our buildings and the energy-consuming things in them are designed and perform, and how we use our buildings and devices, directly and significantly impacts energy use. A poorly insulated home or apartment uses more energy; a well-insulated dwelling uses less. Due to technology and steady improvements in building codes and standards, newer buildings or newly remodeled buildings will outperform older ones in their original condition. The amount of conditioned space that a person occupies is generally proportional to energy use—the bigger the house, the bigger the electric and/or gas bill. But, of course, some very small homes and apartments are very inefficient. Direct and attributable carbon emissions can be volumetrically higher in dense urban environments while simultaneously being significantly lower on a per-capita basis as compared to suburban or rural areas.

As already suggested, building codes and standards are the primary legal and regulatory vehicles for addressing the ways in which energy use in buildings and resultant climate impacts are addressed.[11] Building turnover—the destruction and replacement or remodeling of buildings—is a major driver of building energy performance improvements because most codes and standards apply when a new building is built or when a permit for remodeling is issued. Codes are usually designed and implemented in such a way as to be incrementally more stringent with each passing year or code cycle. As a result, debates about implementation costs and building affordability are always at the forefront of the issues considered when new code versions are considered for adoption.

Similarly, appliance and equipment efficiency standards have been demonstrated to be effective in reducing energy consumption and related emissions associated with air conditioners, heaters and furnaces, refrigerators, clothes washers, toilets and plumbing fixtures[12], and other equipment. Notwithstanding the climate impact efficacy of such standards, debates have often surrounded the tightening and even the existence of these measures.

Over the past several decades, three additional major drivers for improved building energy performance have emerged: utility-administered energy efficiency programs, voluntary green building standards, and a drive toward the

---

[11]   For more information on the built environment, planning, and building codes, *see* Chapter 4, *Adaption to Climate Change at the Subnational Level*, by Shelby D. Green.

[12]   Moving and treating water uses an immense amount of energy. One estimate is that water related energy use accounts for 13% of US electricity use – equivalent to the output of ISO large coal-fired power plants *see* Bevan Griffiths – Sattenspiel and Wendy Wilson, *The Carbon Footprint of Water*, River Network (2009) https://rivernetwork.org/resource-library/carbon-footprint-water

electrification of all energy uses in order to support a shift to renewable energy generation as the source for all energy used in buildings and transportation.

1 Energy efficiency programs for energy use in buildings

Using energy directly or indirectly creates climate changing emissions if the source of that energy is fossil fuels. So using less energy through conservation or more efficient use reduces greenhouse gas emissions. In addition, because power lines and pipelines have resistive and leakage losses, respectively, end-use efficiency improvements yield benefits that compound throughout the system. The regulatory control that state energy commissions exercise over utilities and the ubiquitous reach of utilities with exclusive service territory franchises makes utilities an ideal entity for delivering energy efficiency—except for the fact that efficiency reduces throughput.

A major challenge for regulators and utilities lies in recognizing the resource value of efficiency—as a substitute for distributing and selling commodity electricity or gas—and in appropriately quantifying the greenhouse gas reductions benefits and economic value of efficiency in order to properly size energy efficiency program offerings. In addition, because utilities are typically allowed to distribute (or "socialize") the costs of energy efficiency programs among all customers through rates, program design must address economic free drivers and free riders, that is, the equitable distribution of program benefits and costs.

Some utilities have resisted enhanced efficiency program requirements because of a general trend in developed countries of slow, flat, or even declining load growth over the past several decades. Major regulatory debates have ensued about breaking or "decoupling" the link between utility throughput—and hence greenhouse gas emissions—and utility profits. Approaches being explored by regulators and utilities include revenue and profit adjustments for sales lost due to efficiency program success and, on a broader scale, making an increasing and even dominant share of utility profits dependent on meeting performance standards and not just throughput. The experience of more than five decades of efficiency program success in the US is that efficiency is a great way to reduce greenhouse gas emissions from fossil-burning power plants, so increasing focus on climate-friendly and more efficient use of utility services has strong proponents.

2 Voluntary green power

Another major driver of change in how electricity is generated has been driven through voluntary market offerings of green power, such as electricity generated from renewable resources, or service products created from the bundling of conventional energy with RECs. Many utilities today offer premium-based subscription programs that match renewable energy purchases with customer demand for energy. In this way, customers enact an RPS for themselves, and cumulatively create demand for new renewable energy development. Both utilities and non-utility businesses can also become REC resellers that purchase RECs on behalf of customers seeking to offset the impacts of traditional energy usage. Large corporations, universities, and several governmental agencies have been leaders in voluntarily reducing their carbon footprint through the purchase of RECs. The attractiveness of REC-based products is that the size of the purchase can be scaled to almost any budget, and that the spending can be directed to the least expensive or most desirable forms or locations of renewable energy generation.

3   Electrification

For most of the modern industrial age, energy resources could be divided according to energy density (useful energy per kilogram) and energy storability. The pursuit of high-density energy resources drove the growth of the coal and oil sectors, and eventually, the development of nuclear power plants. Systems that relied on less dense resources—like wind and sunlight—were limited by the costs of the technologies and the efficiency with which they convert diffuse energy into more useful and concentrated forms.

The inability to economically store one of the most useful forms of energy—electricity—was a severe limitation on the application of electricity to the transportation and industrial sectors.

Things are changing rapidly. Discovery and production technology improvements and massive expansion in processing and delivery infrastructure made even diffuse methane into a cost-effective resource. Efficiency in use also helps less dense energy resources compete economically. And battery costs are plummeting. A solar and battery-powered traffic crossing signal is cost-effective because it can use very efficient light-emitting diode (LED) lights and other semiconductor-based hardware, and because expensive connections to the electric grid can be avoided entirely. Such a system also has inherent resilience—it operates even when the grid is down.

Electric personal and commercial transportation is a rapidly growing segment of the market. Of course, the existing stock of fossil-fuel modes of transportation is immense. Many EV drivers value the environmental benefits of switching from petroleum to electricity—benefits that can be increased through voluntary green power programs or installation of solar systems on the home. The health and climate costs of methane gas use are growing in salience as coal generation recedes in more developed countries. Previously uneconomic electric heating alternatives to direct methane gas combustion for heat have become increasingly affordable as technologies like heat pumps have improved.

All these trends have merged with a growing sense of climate urgency to lead increasing numbers of policy advocates and quite a few legislators to call for society to "electrify everything" and to do so with renewable energy. Technical, economic, and social challenges remain, especially in achieving the "everything" part of the goal. Still, falling prices for electricity storage mean that a great deal of electrification can be accomplished today at affordable costs.

C   The Regulation of Energy in the Transportation Sector

In the case of transportation, energy law includes regulations governing tailpipe emissions, regular certifications that vehicles meet emissions requirements, and incentives to encourage consumer adoption of low- or zero-emissions vehicles. This chapter focuses on light-duty (passenger) vehicles—although governments are increasingly becoming aware of, and taking actions to control, emissions from other vehicles (buses, rail, airplanes, heavy-duty, and agricultural vehicles), the majority of transportation-related regulations target passenger cars.

Transportation emissions come from burning petroleum products like gasoline, diesel fuel, bunker fuel, jet fuel, and compressed methane gas. These energy-dense and extremely portable fuels mean that nearly everyone on the planet participates in climate change almost every time they travel for pleasure or work or to move goods.

Energy law is used to impact carbon emissions from transportation sources generally only in indirect ways. The US has implemented seasonally varying requirements for blending ethanol derived from corn into gasoline to address air pollution issues (and support agriculture) but not to address climate change. Research and development programs support alternative fuels development. Tax credits improve the economics of transitioning from petroleum to electricity, compressed methane, or biodiesel fuels.

Fuel economy standards are targeted at reducing criteria pollutant emissions, with an incidental benefit of reducing greenhouse gas emissions.[13] Car-pool lanes and mass transit development and subsidies also yield climate benefits even though they are typically targeted at reducing traffic congestion and urban air pollution. Urban planning aims to make cities and suburbs more friendly to walking and cycling and to interface more effectively with mass transit systems. So-called "cash for clunkers" programs aim at getting highly inefficient vehicles off the road, but were designed to give the automobile market

---

[13]   For a further discussion on emissions control from mobile sources, *see* Chapter 2, *Climate Law Primer: Mitigation Approaches*, by Karl S. Coplan.

a sales stimulus. Government and corporate fleet conversions, supported by tax credits and other incentives, involving the switch to lower-carbon or all-electric vehicles, can produce long-term maintenance savings, help achieve carbon reduction goals, and drive down costs.

Short of outlawing petroleum for use as a fuel, an unlikely legal strategy for most jurisdictions, getting carbon out of transportation will likely continue to rely on a medley of legal and economic mechanisms.[14] In the meantime, improvements in technology, especially in batteries and large-scale production of renewable energy for electric transportation, will be the critical elements of a path to decarbonization in the transportation sector.

D   Climate Change, Extreme Weather, and Energy Security

Universal dependence on energy services—heat, light, refrigeration, motive power, cooling, and transportation—also means near-universal vulnerability to catastrophic energy system failures resulting from extreme weather caused by climate change. Energy security—confidence that energy services will be available and affordable when needed—has become a key issue even in the developed countries that have long enjoyed reliable and affordable energy services.

Hurricanes, droughts and heat waves, freezes and blizzards, tornados, floods, and the other extreme weather hallmarks of a rapidly changing global climate are a threat to energy infrastructure integrity. Vulnerable energy infrastructure includes large and small electric transmission lines, substations, fuel storage facilities, and generation plants; methane gas pipelines, compressors and pumping stations; and production facilities, refineries, pipelines, and distribution and retail sales facilities for petroleum fuels and gas liquids.

Damage to energy infrastructure and reductions in energy system operations has direct effects on consumers and users of energy. Severe weather can lead to loss of life and damage to property, often with the greatest impacts on the poor and communities of color, for whom energy services may be of lesser quality and resilience.

Energy systems damages and failures also impact related critical infrastructure systems. Electric system failures can mean loss of water pumping and

---

[14] However, note that the states of California and Massachusetts have banned the sale of new gas cars after 2035. Brad Plumer and Jill Cowan, *California Plans to Ban Sales of New Gas-Powered Cars in 15 Years*, NEW YORK TIMES (Sept. 23, 2020), https://www.nytimes.com/2020/09/23/climate/california-ban-gas-cars.html; Roberto Baldwin, *Massachusetts to Ban Sale of New Gas-Powered Cars by 2035*, CAR AND DRIVER (Dec. 31, 2020), https://www.caranddriver.com/news/a35104768/massachusetts-ban-new-gas-cars-2035/.

treatment systems, prevent stores from opening, impair first-responder operations, shut down mass transit systems, and lead to all kinds of government services curtailment or interruption.

Not surprisingly, the issue of resilience—the ability of people and systems to make it through such events and to return to normal conditions and operations—is rising in salience and importance. Much of today's energy infrastructure was simply not built for the weather conditions that climate change is creating.

Real-world examples are already available for evaluation. In California, wildfires that many believe are both more common and severe as a result of climate change have also raised questions about the prudent level of investment in climate-hardened energy infrastructure. In Texas, a severe winter event in 2021 led to $50 billion (US) in increased energy costs over just a 70-hour period—several times more than the average annual electricity spending. In Puerto Rico, a massive hurricane almost completely wiped out the electric grid. In every case, the questions, both political and economic, are whether anything or enough was done to prevent the damages from being as severe. And among some commentators, the vulnerability of existing systems has led to calls for a bottom–up reimagining of the structural and market models for providing energy services. The economic consequences of these events put a strain on existing economies and create a mortgage of financial costs that can compete with other critical societal needs.

E   The Growing Trend Toward Full-Economy Climate and Energy Laws

Due to the widespread dependence on fossil energy in every sector of the economy, states seeking to dramatically reduce greenhouse gas emissions have a number of opportunities and challenges. While political conservatives express concern over the economic costs of reducing fossil energy use and transitioning to more sustainable energy resources, green economy and "green new deal" champions argue that system-wide transformation is less expensive than a piece-meal approach due to synergies between and among energy resources. For example, a solar system that generates more electricity than needed in a household during the day can store excess energy in a battery or find use in an electric vehicle, or it can travel to the grid where it can serve other demand and reduce upstream fossil fuel-related emissions. Energy management technologies can enable customers and utilities to reduce energy use during periods of peak carbon emissions, saving money for customers and the utility as well.

Perhaps one of the most interesting developments in climate law and policy has been an expanded focus on aiming for economy-wide carbon emissions reductions. As of the end of 2019, action on policies aimed at a transition to 100% clean energy had been taken in 13 US states: Washington, California, Nevada, Colorado, New Mexico, Wisconsin, Maine, New York, Rhode Island, Connecticut, New Jersey, Hawaii, and Virginia.[15]

A notable example is the state of Virginia, which enacted the Virginia Clean Economy Act (VCEA) in 2020.[16] The VCEA includes a renewable portfolio standard with a tradable RECs program, support for small-scale renewable energy resources, state participation in a regional greenhouse gas emissions reduction program, energy efficiency standards, and strong support for emerging clean energy technologies like offshore wind generation and energy storage. One of the most notable features of the VCEA is an explicit focus on energy justice and equity. The act steers incentives toward clean energy development in environmentally disadvantaged communities and emphasizes clean energy job creation.

The goal of the VCEA is zero carbon emissions from the electricity sector by the year 2050. Still to come will be policy efforts to address building energy use, direct industrial use of energy, and the transportation sector.

CONCLUSION

The regulation of greenhouse gas emissions from the energy sector involves a wide array of industries, processes, governing agencies, and subject matters, from foundational economics and regulatory constructs to power generation, building electrification, gas decarbonization, and transportation. The future of regulating energy emissions must—and increasingly does—include energy and environmental justice goals and programs, and more engagement from frontline communities and those most directly impacted by fossil fuel production and use. In addition, more and more jurisdictions are adopting sweeping climate and energy legislation, mandating zero-carbon electricity produc-

---

[15]   Sam Ricketts et al., *States Are Laying a Road Map for Climate Leadership*, CENTER FOR AM. PROGRESS (Apr. 30, 2020), https://www.americanprogress.org/issues/green/reports/2020/04/30/484163/states-laying-road-map-climate-leadership/.   *See also*, *Clean Energy Policy Tracker*, NATIONAL REGULATORY RESEARCH INSTITUTE, https://www.naruc.org/nrri/nrri-activities/clean-energy-tracker/.

[16]   Alena Yarmosky, *Governor Northam Signs Clean Energy Legislation*, OFFICE OF THE VIRGINIA GOVERNOR (Apr. 12, 2020), https://www.governor.virginia.gov/newsroom/all-releases/2020/april/headline-856056-en.html. The VCEA is in part modeled on the New York Climate Leadership and Community Protection Act, which also includes environmental justice provisions, large-scale renewables and energy efficiency requirements, and economy-wide decarbonization goals. *See*, New York's Climate Leadership and Community Protection Act (CLCPA), https://climate.ny.gov (last visited Apr. 18, 2021).

tion, and targeting zero-emissions economies with mid-century mandates. Regulating emissions from the energy sector will continue to require aggressive, multifaceted action across a range of industries and initiatives.

4.   Adaption to climate change at the subnational level

Shelby D. Green

I   THE PORTENTS OF CLIMATE CHANGE

When we finally looked up a few decades ago, we saw a climate disaster approaching. At least that is what scientists are predicting, given the world we have made. We built too close to the sea. Nearly two-thirds of urban areas are in low elevation coastal zones or located along flood-prone rivers. Sea-level rise and storm surges could result in the eventual abandonment of some urban districts.[1] We built on lots that were too large and too far from work, causing us to put too many cars with their noxious fumes and noise on the highways. We took down too many trees, taking away their natural capacity to reduce carbon and provide shade and cooling.

Changes linked to the rapid increases in atmospheric greenhouses gases are already affecting global climates and the oceans, and these changes are expected to accelerate in the coming decades. The most widely felt change is warming temperatures, not only of the air but also oceanic surface waters. These temperature changes set in motion a chain reaction that first drives more precipitation, and then the patterns of storms and ocean currents. Changes in river flows follow, and then sea levels rise from the natural expansion of water as it warms and increases the volume of water driven by the melting of large ice caps.[2]

The magnitude of adverse consequences from climate change is indeed frightening: rising sea levels of up to a meter in the next century; sea levels have already risen by 20 centimeters or more in many areas, and rise rates

---

[1]   During Superstorm Sandy, the storm knocked out transformers, leaving Manhattan without power; the Empire State Building went dark. There was no cell service. Subways flooded with hundreds of thousands of tons of debris clogging pathways.

[2]   IPCC Fifth Assessment Report, https://www.ipcc.ch/assessment-report/ar5/ [hereinafter IPCC].

appear to be accelerating, although there is no consensus as to how fast such changes will be in the future, as estimates of global sea-level rise may not reveal changes at local levels, which are driven by local conditions. Where seas are rising, the threat to coastal lands is not just gradual inundation of low-lying areas but also of increased erosion rates even along higher shorelines. These will lead to more impactful coastal storm events, with fierce winds and saltwater intrusion into the groundwater.

Paradoxically, there will be both floods and drought. California recently experienced both.[3] Deluges from massive amounts of precipitation (as much as 10% by the 2020s and 21% by the 2080s) will lead to ocean and riverine flooding. Some parts of the country will seem perpetually under-water. The New York City floodplain could expand 23% in the 2020s and, by the 2050s, to cover a quarter of the city.[4]

Too much water in some places and too little in others will lead to all kinds of weird results, like water rationing and lower crop yields. Dry forests will lead to an increased risk of wildfires. While increasing concentrations of carbon dioxide may positively affect growth rates of coastal vegetation, the impact of ocean acidification is also predicted to be severe on coral reefs and other calcifying organisms.

The expected disruptions to life can be described in terms of more this and more that: more costs (for energy, for cooling, for delivering water, for mending roads, for fire suppression, for emergency services, for insurance), more illness (from heat, pests, noxious plants), more pollution (in the air, the rivers, and the ocean), more dislocations (loss of homes and livelihoods),[5]

---

[3]   In 2015, the State of California declared a state of emergency on account of drought. The executive order called for, among other things, a 25% statewide reduction in water consumption. Exec. Order No. B-29-15 (2015),
https://www.ca.gov/archive/gov39/wp-content/uploads/2017/09/4.1.15_Executive _Order.pdf; *see also* http://ca.gov/drought/.
[4]   *New York City's Risk Landscape: A Guide to Hazard Mitigation*, NYC Emergency Management (Nov. 2014), https://www1.nyc.gov/assets/em/downloads/ pdf/hazard_mitigation/nycs_risk_landscape_chapter_4.3_flooding.pdf. Yet, cities continue to encourage development in flood plains, including the 17 million square feet or 26-acre Hudson Yards development on the West Side of New York, which is within the 100-year flood zone. The developers insist that much of the construction occurred on a platform 40 feet above sea level and is designed to resist flooding—the platform puts the first floor above the floodplain and electrical and support systems will be above ground. Jim Dwyer, *Still Building at the Edge of the City, Even as Tides Rise*, New York Times, December 4, 2012; *see also* Kyle Chayka, *Developers Keep Building in Sandy Flood Zones*, New York (October 2, 2015).
[5]   Thousands of lives were lost from Hurricane Katrina and Superstorm Sandy, scores of communities were destroyed, and economic losses were in the billions.

more damage to beaches, more loss of cultural sites (statues inundated and historical developments washed away).

Who is impacted? Everyone. However, some will suffer more than others, such as individuals who cannot afford sturdy housing or must labor out of doors; also, cities with their micro-climates,[6] and because of the interconnectedness of infrastructure (both natural and man-made) that transport people (roads, bridges, and rail lines), water (pipes and pump stations), waste (sewage treatment plants), and light (power plants), as well as networks that facilitate living and transacting business, such as telephone communications for banking and emergency services. Our older cities' infrastructure is ailing, fragile, and deficient; it will eventually fail to support growing urban populations—hundred-year-old pipes were designed to serve a fraction of the population they currently serve. And, when one system fails, there is a cascading effect on other systems: No power means no food is delivered, no surgeries are performed, no subway trains operate.[7]

More than half the world's population and more than 80% of the United States population live in urban areas. And this level is increasing.[8] Population density and development lead to the urban heat island effect—another attribute of the urban micro-climate—occurring when naturally vegetated surfaces are replaced with impervious surfaces that absorb, retain, and reradiate more solar energy than grass and trees. As average air temperatures rise, so does the urban heat island effect.

II   TAKING ACTION: MITIGATION AND ADAPTATION

Climate change is a vagabond, going everywhere when it chooses. It is omnipotent and omnipresent. As the prevailing view is that climate change is human

---

[6]   These are locales with distinct climate conditions due to urban development that reveals varying environmental conditions, like temperature, light, wind speed, and moisture.

[7]   On August 8, 2007, an intense rainfall and thunderstorm event in New York City during the morning commute dumped between 1.4 and 3.5 inches of rain within two hours. This rainfall resulted in a cascade of transit system failures—eventually stranding 2.5 million riders, shutting down much of the subway system, and severely disrupting the city's bus system.

[8]   Since 2000, many major cities have increased their new home construction share, while regional levels have declined. In 2008, the city of Portland issued 28% of all building permits compared to 9% in the region. In Denver, that level was 32%, compared to 5% in the region. In Sacramento, 27% compared to 9% in the region. New York City issued 63% of all building permits, and Chicago issued 45%.

caused,[9] reversing it may be beyond human powers, at least in the short term. There is nowhere to run. Although we cannot stop the fierce blizzards or ravaging hurricanes, we can work to reduce the conditions giving rise to them and their ferocity,[10] to halt the progression of climate change and adapt. We can act to enable resistance and resilience to the present and impending changes.

Is it all anthropomorphic? Whether or not it is, the world is responding to climate change portents by both mitigation efforts and adaptation measures. Mitigation primarily focuses on cutting greenhouse gas emissions, thought to be the main culprit in climate change. It is both scientific and political because it embraces new levels of regulation of industry and private conduct, requiring radical changes in the way the world operates, its productivity goals, and industrial investments. Mitigation is reflected in national policy: calling for efficient energy production and use; providing funding for research, development, and investment in new energy technologies in their operations; requiring a retooling of public buildings for efficiency; adopting a national energy code; and requiring recipients of federal funds to introduce efficiency technologies.[11]

Adaptation means actions to take at the individual, local, regional, and national levels to manage and reduce risks from changed climate conditions and prepare for impacts from additional changes projected for the future. It aims to reduce the impact of climate stresses on humans and natural systems. It consists of a multitude of behavioral, structural, and technological measures. Adaptation strategies can have many dimensions: timing (proactive or reactive); scope (local or regional or global); purposefulness (autonomous or planned); and agency (private or public). It is often described in three strategies: protect, accommodate, and retreat. Protection measures feature hard armoring of the shores against surges and sea-level rises, with barriers (seawalls, levees, automatic gates) that open and close in response to sea-level rises (like the Venice Lagoon and measures proposed for New York City).

---

[9] Daniel A. Bader et al., *Urban Climate Science*, Cambridge University Press (2018), at 28,
https://uccrn.ei.columbia.edu/sites/default/files/content/pubs/ARC3.2-PDF-Chapter-2-Urban-Climate-Science-compressed.pdf.

[10] *See* Shelby D. Green, *Zoning Neighborhoods for Resilience: Drivers, Tools & Impacts*, 28 Fordham Envt'l L. Rev. 41 (2016); Karl S. Coplan, LIVE SUSTAINABLY NOW: A LOW-CARBON VISION OF THE GOOD LIFE (2019); Michael B. Gerrard and John C. Dernbach, LEGAL PATHWAYS TO DEEP DECARBONIZATION IN THE UNITED STATES.

[11] *See, e.g.,* Aspen Climate Action Plan, https://www.cityofaspen.com/DocumentCenter/View/4506/Aspens-Climate-Action-Plan- (describing the city's ambitious greenhouse gas (GHG) reduction targets, to be achieved through, *inter alia*, increased use of renewable energy sources, facilitating alternative transportation methods, retrofitting homes with efficient equipment and technologies, and reducing waste through composting and recycling).

Accommodation measures recognize that the forces of climate change cannot be resisted, but that peace must be made with nature. Rising heat levels must be ameliorated by efficient technologies for cooling; ferocious winds must be resisted by fortified structures; homes must be set back to avoid storm surges; drought must be managed by heat and drought-resistant crops and water use efficiency; floods must be repelled by barriers, building design, and location. There must be planning, locally and regionally, for effective evacuation in storm events. Public health measures must address changes in disease vectors. Zoning must include green spaces.

Retreat measures contemplate relocation of vulnerable populations and communities by demolishing vulnerable structures, sometimes using eminent domain to acquire properties in vulnerable areas. Other measures consider zoning land as undevelopable, imposing conservation easements, and requiring setbacks and elevations (of sites and structures).[12]

Besides averting individual harm and loss, adaptation promises many societal benefits, both short term and long term. Growing urban forests can bring beauty as well as shading to reduce energy consumption. Restoring wetlands can provide valuable habitat for fish and wildlife, as well as flood protection to nearby communities. Conserving mangrove ecosystems can protect coastal communities from fierce storm systems and serve to absorb carbon from the atmosphere at the same time. Efficacious adaptation requires forward-looking and integrated planning, and they may sometimes mean sizable costs and diversions of resources.

A   Actions on All Fronts

Heeding the portents, scientists, advocates, politicians, and governments have gone into action. As climate change is a global phenomenon, nations have been working to combat it through international conventions. The global march toward mitigation efforts has remained largely in sync—essentially controlling greenhouse gas emissions.[13] For a while, there was an acknowl-

---

[12]   *See generally,* Robert R.M. Verchick & Lynsey R. Johnson, *When Retreat is the Best Option: Flood Insurance After Biggert-Waters and Other Climate Change Puzzles,* 47 J. Marshall L. Rev. 695, 697 (2013) (explaining that retreat involves the removal of people and property and restricting development in existing communities).

[13]   The Paris Agreement was adopted on December 12, 2015,
https://unfccc.int/resource/docs/2015/cop21/eng/l09.pdf. It recognized that climate change is a common concern of humankind. Article 2(2) states that "adaptation is a global challenge faced by all, with local, subnational, national, regional and international dimensions." Article 7(1) establishes a "global goal on adaptation of enhancing adaptive capacity, strengthening resilience and reducing vulnerability to climate change, with a view to contributing to sustainable development and ensuring an ade-

edgment of climate change in the United States executive offices. Hence, all kinds of federal initiatives were adopted to respond through both mitigation and adaptation measures.[14] The previous administration questioned the reality of climate change phenomena and had taken concrete steps to frustrate efforts at mitigation—withdrawing from global pacts on climate change; relaxing environmental rules aimed at controlling pollution; abandoning the clean power coal plant initiative, and instead working to prop up the coal industry. This meant that the efforts to arrest its progress, it seems, had to be taken at the subnational level, and in the United States, that meant by states and local jurisdictions. Efforts by this cohort are particularly urgent where federal action had waned or had become too political and non-existent.

Subnational governments are uniquely able to develop policies, share and coordinate their efforts with other levels of government. They are often at the forefront of developing priorities and putting plans into action. Subnational governments often serve as incubators for innovative actions and policies. Neighborhoods experience micro-climates—locales with distinct climate conditions[15]—such that they require systems calibrated to address the unique challenges and the range of disaster risks manifested from these local climate

---

quate adaptation response in the context of the temperature goal referred to in Article 2" [well below 2°C above pre-industrial levels and pursuing efforts to limit the temperature increases to 1.5°C above pre-industrial levels]. (Article 2(1)(a)); *see also* http://www.economist.com/news/international/21684144-what-expect-after-deal-exceeded-expectations-green-light (reporting that more than 195 countries attending the meeting agreed to the stated goals). For an interesting treatment of what cities can do toward reducing greenhouse gases, *see* Jonathan Rosenbloom, REMARKABLE CITIES AND THE FIGHT AGAINST CLIMATE CHANGE: 43 RECOMMENDATIONS TO REDUCE GREENHOUSE GASES AND THE COMMUNITIES THAT HAVE ADOPTED THEM (2020).

[14] For a summary of such measures, *see* Shelby D. Green, *Building Resilient Communities in the Wake of Climate Change While Keeping Affordable Housing Safe from Sea Changes in Nature and Policy*, 54 Washburn L. J. 554–45 (2015); Shelby D. Green, *Zoning Neighborhoods for Resilience: Drivers, Tools & Impacts*, 28 Fordham Envt'l L. Rev. 41 (2016).

[15] *New York City Microclimate Policy Applying Green Infrastructure to Mitigate Environmental Health Impacts caused by the Urban Heat Island Effect and Heat Waves: A Platform for Climate Change Resiliency in New York City* 3 (2012). "Microclimates [may be] created naturally by geographical changes in the environment such as coastal zones, topographical differences in altitude, and [by] manmade environments." *Id.* An urban micro-climate is said to refer to discrete areas, whereas a consequence of urban development, environmental conditions vary from those in nearby regions. The environmental variations include temperature, light, wind speed, and moisture. R. Geiger, The Climate Near the Ground 1965; *see also* Evyatar Erell, David Pearlmutter, and Terrence Williamson, URBAN MICRO CLIMATE: DESIGNING THE SPACE BETWEEN BUILDINGS 15–17 (2012); Microclimates, National Meteorological Library and Archive Fact sheet 14 — MICROCLIMATES (*version 01*), www.metoffice.gov

conditions. Because the vulnerability to climate change's effects will vary by location, the degree of development, and demographic factors (where and how people live), adaptation efforts need to be focused on the micro-level within cities. The approaches discussed below have included comprehensive planning, regulatory mandates, economic incentives, and educational programs to encourage wise stewardship.

III   COMMITMENT TO MITIGATION ON THE STATE AND LOCAL LEVELS

As the United States formally joined the Paris Agreement, so did state and local governments pledge their commitment to the goals of that international agreement through various coalitions and alliances. The *Under2 Coalition* members committed to reduce GHG emissions 80–95% below 1990 levels by 2050, or to limit emissions to 2 metric tons of CO_2e per capita annually. The *Global Covenant of Mayors* pledged to work alongside nations in the same vein. The *US Climate Alliance*, comprising 14 states and Puerto Rico, has committed to meeting their share of the United States' nationally determined contribution under the Paris Agreement (a 26–28% reduction in GHG emissions below 2005 levels by 2025). The *US Climate Mayors* consists of nearly 400 cities and is committed to intensifying efforts to meet climate goals. Prompted by the United States withdrawal from the Paris Agreement by former President Trump, *We are Still In*, consisting of leaders from America's state and city governments, as well as corporations, declared its support for the Paris Agreement.[16] More than 1,055 municipalities from all 50 states have signed the US Mayors Climate Protection Agreement, and many of these communities are actively implementing strategies to reduce their emissions.[17]

A   GHG Reduction Targets

Outside of and consistent with promises under coalitions and alliances, states and cities are designing and implementing climate mitigation measures in multifaceted and inventive schemes. More than 20 states and over 100 cities have enacted GHG reduction targets—some with ambitious goals, like reduc-

---

.uk/learning/library/publications/factsheets; *see also* Cornell Gardening Resources—Microclimates, http://www.gardening.cornell.edu/weather/microcli.html.

[16]   *America's Pledge (Phase 1 Report): States, Cities, and Businesses in the United States are Stepping Up on Climate Action* [hereinafter America's Pledge].

        https://www.bbhub.io/dotorg/sites/28/2017/11/AmericasPledgePhaseOn eReportWeb.pdf at 14, 29–30

[17]   https://nca2014.globalchange.gov/highlights/report-findings/responses.

ing GHG emissions by half by the next decade.[18] Under the Carbon Neutral Cities Alliance, cities pledge to reduce GHG emissions by 80% by 2050.[19] The Portland, Oregon Climate Action Plan establishes a goal to reduce GHG emissions by 53% below 2006 levels by 2030 and promises to be carbon-free by the year 2050.[20] The Minneapolis, Minnesota Climate Action Plan projects a goal of 30% reduction below 2006 levels by 2025, and under its Climate Action and Resilience Plan, sets 2050 as the year when it will be carbon neutral.

The states and cities aim to meet these targets through decreased reliance on fossil fuels, greater exploitation of clean energy and energy conservation.[21] Already, more than 70 cities and counties are powered by 100% clean energy.[22] Austin, Texas has powered all municipal buildings with 100% renewable energy since 2011 and has also set goals to reduce energy consumption in buildings by 5% annually. That city's Resource, Generation and Climate Protection Plan aims to fill at least 55% of customer demand from renewable energy resources by 2025 and 65% by the end of 2027. Boston is on track to achieve its goal of carbon neutrality in government operations by 2050 and has installed solar energy systems on municipal facilities through the Renew Boston Trust. California has 66 cities and counties at 100% renewable energy. San José has a goal to source 100% of its power mix from non-hydro-renewables by 2050 and 60% by 2030.[23]

More than half the states have adopted mandatory renewable portfolio standards (RPS) and a handful of others have voluntary renewable energy goals, most prominently, solar energy.[24] Solar energy is supported through

---

[18]   *See* GHG Emission Targets: Center for Climate and Energy Solutions, "Climate Programs and Policy Maps," https://www.c2es.org/us-states-regions#states.

[19]   http://climateinitiativesplatform.org/index.php/Carbon_Neutral_Cities _Alliance; http:/ /www2.minneapolismn.gov/www/groups/public/@citycoordinator/ documents/webcontent/wcmsp-194790.pdf.

[20]   In the city's Climate Emergency Declaration (June 30, 2020), it adopted a new target to achieve a 50% reduction in carbon emissions from 1990 levels by 2030, and reach net-zero carbon emissions before 2050. City bureaus are responsible for implementing actions to help meet this commitment. https://www.portlandoregon.gov/water/ 63676; *see also* Green Cincinnati Plan 2013, (planned net-zero police station).

[21]   Progress Toward 100% Clean Energy In Cities & States Across The U.S., 100-Clean-Energy-Progress-Report-UCLA-2.pdf. [hereinafter Progress].

[22]   Progress, *supra* note 21, at 2.

[23]   David Ribeiro, Stefen Samarripas, Kate Tanabe, Alexander Jarrah, Hannah Bastian, Ariel Drehobl, Shruti Vaidyanathan, Emma Cooper, Ben Jennings, and Nick Henner, *The 2020 City Clean Energy Scorecard, ACEE*. October 2020 I Report U2008 [hereinafter Score Card] at 22.

[24]   Cal. Pub. Resources Code §25980 (striving to balance the interest in solar panels against shade from trees); City of Minneapolis, Minneapolis Climate Adaptation Plan, http://www2.minneapolismn.gov/sustainability/climate-action-goals/climate-action

zoning measures that allow the placement of solar panels in places heretofore off-limits;[25] others with cost-sharing and production-based incentives.[26] North Carolina has the second most installed solar capacity in the nation, with approximately 7,000 MW of cumulative renewable energy capacity. In Mississippi and Alabama solar markets are growing at a fast pace.[27] Portland, Oregon has deployed onsite solar installations and battery storage at municipal facilities.[28]

Other forms of alternative energy sources being supported by local governments include geothermal energy (which is obtained by harnessing heat from the ground to both heat and cool homes).[29] In Texas, wind is the largest source of power.[30] Massachusetts has created requirements for offshore wind as well as new solar procurement programs.[31] Wind turbines can provide up to 30% of the energy consumed by a household. However, they are subject to regulations and technical specifications, such as the maximum distance at which the facility is located from the place of consumption, and the power required and permitted for each property. Zoning amendments inform plans for the erection of wind turbines.[32] For an equitable and efficient distribution of energy, smart grids are being created.[33] Vermont's climate action plan encompasses the goals outlined in the state's Comprehensive Energy Plan. Key aspects include accelerating the adoption of advanced wood heat to replace high-emitting systems (aiming to reach 30% of Vermont's thermal needs by 2025), strengthening the

---

-plan; City of Phoenix, Ariz., *City adopts new goal to reduce greenhouse gas (GHG) emissions from city operations to 15 percent by 2015*, https://www.phoenix.gov/oep/environment/climate. Two types of solar panels generate heat into electricity. Thermal solar panels reduce or eliminate the consumption of gas and diesel, and reduce CO2 emissions. Photovoltaic panels convert solar radiation into an electric current that can power any appliance. This is a more complex technology and is generally more expensive to manufacture than thermal panels.

[25]   The City of Hartford, Connecticut allows free-standing solar panels on historic properties.   https://hartfordclimate.files.wordpress.com/2016/12/historic-properties -guidelines.pdf.

[26]   http://www2.minneapolismn.gov/sustainability/buildings-energy/solar.

[27]   America's Pledge, *supra* note 16, at 37.

[28]   PROGRESS TOWARD 100% CLEAN ENERGY IN CITIES & STATES ACROSS THE U.S., 100-Clean-Energy-Progress-Report-UCLA-2.pdf.

[29]   For some resources on alternative energy, *see Center for Sustainable Systems, Univ. of Mich.,* http://css.umich.edu/sites/default/files/Geothermal%20Energy_CSS10 -10_e2019.pdf.

[30]   America's Pledge, *supra* note 16, at 37.

[31]   America's Pledge, *supra* note 16, at 42.

[32]   *See Distributed Wind Energy: Zoning & Permitting: A Toolkit for Local Governments*, https://www.cesa. org/wp-content/uploads/Distributed-Wind-Toolkit. pdf.

[33]   Green Cincinnati Plan 2013 (planned net-zero police station).

used electric vehicle market, and supporting free legal services for new climate economy entrepreneurs.[34]

B   Carbon Pricing

Ten states have adopted carbon pricing regulations to reduce emissions, the first being California.[35] Nine other states have followed under the Regional Greenhouse Gas Initiative (RGGI) with commitments to create a market-based system that sets a cap on emissions from the electric sector. Under the program, Connecticut, Delaware, Maine, Maryland, Massachusetts, New Hampshire, New York, Rhode Island, and Vermont have reduced CO_2 by more than 45% since 2005.[36]

C   Transportation

Cities and states have created programs to decrease transportation energy use through transportation planning support for cleaner vehicles. The plans include goals for reducing vehicle miles traveled (VMT) or GHG emissions from transportation and for increasing the proportion of trips taken using alternative modes of transportation. Reaching these goals depends on location-efficient zoning and planning cities so that residents can access major destinations by multiple transportation modes. Portland's 2035 Transportation System Plan includes designs for transportation corridors and incentives for use of efficient vehicles as well as electric vehicle charging infrastructure.[37] The St. Paul 2040 Comprehensive Plan in 2019 set a goal to decrease VMT by 40% by 2040.[38]

California has led the nation on GHG standards for motor vehicles, more stringent than federal regulations, and more than a dozen other states have followed.[39] Even though the numbers and variety of high-efficiency, low-emission vehicle options have increased, states continue to push the indus-

---

[34]   https://www.c2es.org/site/assets/uploads/2018/11/VT_2017_Preliminary_Plan.pdf;
https://outside.vermont.gov/sov/webservices/Shared%20Documents/2016CEP_Final.pdf

[35]   California's Global Warming Solutions Act of 2006, AB 32; America's Pledge, *supra* note 16 at 40.

[36]   America's Pledge, *supra* note 16, at 41–42.

[37]   Score Card, *supra* note 23, at 31.

[38]   More than 30 states have strategies to improve access to "multimodal" freight transport. In the District of Columbia, the Clean Energy DC Omnibus Amendment Act of 2018 was an early adopter of high-impact transportation efficiency strategies, like increasing freight system efficiency. America's Pledge, *supra* note 16 at 43.

[39]   America's Pledge, *supra* note 16, at 42.

try by requiring an increasing percentage of an automaker's sales in the state to be zero-emission vehicles (ZEV)[40] and more than half of state governments have policies requiring state vehicle fleets to become more efficient.[41] By a 2007 executive order on climate action, the city of Boston required municipal departments to purchase hybrid or high-efficiency vehicles, and nearly 15% of Boston's fleet is currently composed of efficient vehicles. Austin, Texas and Portland, Oregon have similar mandates, and in Portland, some 14% of the city's fleet is composed of energy-efficient vehicles including hybrid, plug-in hybrid, and battery electric.[42] As electric vehicles need charging sites, some cities, like Los Angeles, provide incentives for residential and commercial electric vehicle chargers. Some states have amended their laws to preclude condominium associations from banning charging stations on the premises.[43]

A number of cities are evaluating congestion pricing in the urban core as a way to ease traffic stalls and simultaneously generate revenue for more efficient forms of travel.[44] New York State approved the first congestion pricing program in the United States, which is scheduled to go into effect in Manhattan's central business district in 2021.[45]

Many states have created and/or are encouraging car sharing programs,[46] and many communities are acting to encourage biking, through dedicated street lanes and bike pathways.[47] The National Complete Streets Coalition estimates that more than 1,000 cities and counties have adopted policies that aim to improve the safety of their street networks for alternative modes of travel and to reduce congestion.[48]

---

[40]   *Id.* at 42.

[41]   *Id.* at 43.

[42]   Score Card, *supra* note 23, at 36.

[43]   Cal. Civil Code §4745.

[44]   Score Card, *supra* note 23, at 110.

[45]   *Id.*

[46]   Car sharing program: American Council for an Energy Efficient Economy, "City Energy Efficiency Scorecard 2017," http://aceee.org/sites/default/files/publications/researchreports/u1705.pdf;

[47]   Bike sharing program: American Council for an Energy Efficient Economy, "City Energy Efficiency Scorecard 2017," http://aceee.org/sites/default/files/publications/researchreports/u1705.pdf; *see also City of Portland and Multonah County, Climate Action Plan*, 2009, Year Two, Progress Report (April 2012) (calling for a system of bike paths throughout the city).

[48]   "Complete Streets Policies Nationwide," Smart Growth America, accessed September 29, 2017, https://smartgrowthamerica.org/program/national-completestreets -coalition/policy-development/policy-atlas/.

D   Energy Use, Building and Energy Codes, and Efficient Appliances

Almost all states have adopted energy codes for either residential and/or commercial buildings and nearly half have energy efficiency resource standards (EERS) in place.[49] More than half have incentives to help commercial and industrial facilities install combined heat and power systems for greater efficiency. The new codes focus on energy usage, aiming for energy efficiency from a number of different angles, including new construction standards that incorporate energy efficiency.[50] They require new and renovated buildings to substantially reduce the amount of energy they use over their useful lives. Homes built to the 2012 energy code use 50% less energy per square foot than a home constructed in the 1970s.[51] There are two model national energy codes, one for residential buildings and another for commercial buildings. The International Energy Conservation Code (IECC) applies to residential buildings. For commercial buildings, the American Society of Heating, Refrigerating and Air-Conditioning Engineers (ASHRAE) Standard 90.1, developed jointly by ASHRAE and the Illuminating Engineering Society, applies.[52] In 2019, New York City enacted the Climate Mobilization Act,[53] which sets ambitious energy efficiency goals: requiring "covered buildings" (those larger than 25,000 square feet) to reduce building emissions 40% by 2030 and 80% by 2050. New building codes in cities including St. Louis, Austin, and Seattle require newly constructed residential, multifamily, and commercial buildings to be solar ready; in Seattle, if solar is not feasible, buildings must meet energy efficiency savings requirements more stringent than the current code. Other adopted codes require new construction to be electric vehicle ready.[54]

New codes also incorporate design standards, the most common of which is the LEED standard.[55] Miami-Dade requires new county-owned, leased, or

---

[49]   America's Pledge, *supra* note 16, at 39.

[50]   *See* New York State Energy Planning Bd., *The Energy to Lead: 2015 New York State Energy Plan* 18–23, 69–77 (2015), http://energyplan.ny.gov/.

[51]   *See* Charles F. Kutscher, Jeffrey S. Logan, Timothy C. Coburn, *Accelerating the US Clean Energy Transformation: Challenges and Solutions by Sector*, https://www.colorado.edu/rasei/sites/default/files/attached-files/accelerating_the_us_clean_energy_transformation_final.2.pdf.

[52]   Score Card, *supra* note 23 at 65.

[53]   Local Law 97.

[54]   Score Card, *supra* note 23 at 66.

[55]   The United States Green Building Council Leadership in Energy and Environmental Design (LEED) offers a system for implementing and assessing energy efficiency and design through a rating system. www.usgbc.org. The city of Atlanta

managed construction projects to obtain US Green Building's LEED silver certification and remodeling.[56] Miami-Dade offers expedited plan review for registered or certified green buildings according to LEED.[57] Other governments are adopting requirements for reflective coatings, green and cool roofs,[58] which may serve to keep buildings cooler in rising average temperatures. Other governments are supporting energy star qualified homes.[59]

The Minneapolis Climate Action Plan contains a program for passive houses, i.e., homes that are energy self-sufficient, heated and cooled using non-mechanical methods. The plan contains standards for continuous insulation, airtightness, thermal bridging, high-performance windows, and mechanical heat recovery.[60] The use of optimal daylight plays an integral role in passive energy systems, such that buildings are positioned and located to allow and make use of sunlight throughout the whole year. The identified benefits of passive homes are high air quality, even temperature distribution, low noise levels, and reduced heating and cooling costs.[61]

---

requires all city-funded projects over 5,000 square feet or over \$2 million to meet LEED Silver standard. Atlanta Division of Sustainability and U.S. Department of Energy, Atlanta: Power to Change Sustainability Plan Executive Summary 2010–2011, http://clatl.com/images/blogimages/2010/10/26/1288116274-atlsustainplan.pdf.

[56]   County Ordinance 07-65; *see generally* https://www.miamidade.gov/global/economy/resilience/county-climate-programs.page.

[57]   https://www.miamidade.gov/permits/home.asp?cat=build&subcat=plan&filter1 =green; *see also,* City of Minneapolis, Minneapolis Climate Adaptation Plan (2013); City of Phoenix, Ariz., *City adopts new goal to reduce greenhouse gas (GHG) emissions from city operations to 15 percent by 2015,* https://www.phoenix.gov/oep/environment/climate.

[58]   To reduce the urban heat island effect, Chicago will add 6,000 buildings with cool roofs by 2020, which is expected to reduce temperatures by an average of 7 degrees. Chicago Climate Action Plan, at 22. Kansas City, Missouri, has installed over 450,000 square feet of green rooftop from 1999 to 2015. In 2014, Kansas City was ranked on the top 10 U.S. metro areas that experienced intense urban heat islands. The EPA developed a case study to demonstrate the environmental and health impacts of the green roofs in Kansas City. U.S. Environmental Protection Agency. (2018). Estimating the environmental effects of green roofs: A case study in Kansas City, Missouri. EPA 430-S-18-001. www.epa.gov/heat-islands/using-greenroofs-reduce-heat-islands. Among other things, the study found that the green roof systems can retain up to 29 inches of stormwater per year on average, that green roofs save \$41,587 in electricity costs, and that green roofs significantly reduce air pollutant emissions and the urban heat island effect.

[59]   *See Learn how Portfolio Manager helps you save,* ENERGYSTAR, http://www.energystar.gov/buildings/facility-owners-and-managers/existing-buildings/use-portfolio-manager/learn-how-portfolio-manager.

[60]   http://www2.minneapolismn.gov/sustainability/buildings-energy/WCMSP-215225

[61]   *Id.*

Nearly a dozen states have adopted at least one appliance or equipment energy efficiency standard for products not currently covered by standards set by the federal government. California has adopted Time-of-Use rate packages to encourage consumers to use less energy during high demand times and when solar generation is rapidly declining.[62] Seattle's Building Tune-Ups policy requires owners of large commercial buildings to perform energy assessments and tune-ups—a form of retro-commissioning—every five years to optimize the performance of their energy and water systems.[63]

E   Public Infrastructure Converting Streetlights to LEDs

States and cities are reducing governments' energy use in their public infrastructure by installing LED lighting. Over its lifetime, efficient outdoor lighting not only can cut overall energy consumption but can reduce maintenance spending as LED lights are replaced less often. LED technologies can offer savings of 70% relative to traditional light sources. Programming lighting to turn on only during hours needing illumination will also extend lamp lifetimes and save energy.[64]

F   Building Energy Performance Standards and Benchmarking

More cities have adopted benchmarking and transparency policies to serve as foundational steps.[65] This strategy is a crucial step in understanding energy performance and enables cities to identify energy efficiency investment opportunities and track energy savings. The most important type of benchmarking requires owners to measure and report their annual energy usage to the city, which reports are available for public inspection. In Boston, in accordance with its Energy Reporting and Disclosure Ordinance, the city benchmarks 100% of municipal buildings. Under the Chicago Benchmarking Energy Ordinance created the Chicago Energy Rating System, all buildings over 50,000 square feet require an energy performance rating, which owners must post prominently and disclose upon a sale or lease of the property. Washington, D.C., St. Louis, Missouri, and New York have coupled benchmarking with building energy performance standards (BEPS). These standards require owners to

---

[62]   *See generally* Kutscher, *supra* note 51.

[63]   Score Card, *supra* note 23, at 71.

[64]   Score Card, *supra* note 23, at 23, 36.

[65]   *See* generally, Zachary Hart, "The Benefits of Benchmarking Building Performance." Institute for Market Transformation, December 2015, accessed October 23, 2017, http://www.imt.org/uploads/resources/files/PCC_Benefits_of_Benchmarking.pdf.

undergo an energy audit to assess a building's energy consumption to discover areas for energy-saving improvements. St. Louis, Missouri recently adopted BEPS that require large existing buildings to meet a certain standard by 2025 and increasingly stringent standards through subsequent compliance cycles.[66] Some cities require large buildings to undergo retro-commissioning to ensure all building systems are running at optimal energy efficiency.[67]

G   Integrated Programs

Cities are adopting shared distributed energy systems, such as district energy, microgrids, and community solar gardens.[68] A district energy system incorporates combined heat and power and can reduce the amount of energy wasted from 67% to 20%. Microgrids increase efficiency in the generation and distribution of electricity, by shortening the distance to end users, thereby reducing line losses by an annual average of 4–5% compared with the main grid's transmission and distribution system. This also means energy savings as the generation needs are reduced, in some cases producing additional energy savings of 30–40% relative to a traditional generation system. Microgrids can house both renewable energy and fossil fuel resources. Community solar programs are shared solar systems to which customers subscribe and, in some models, receive credit on their utility bill for the amount of clean energy produced by their share.[69]

H   Mitigation of Urban Heat Islands

The urban heat island effect occurs when naturally vegetated surfaces are replaced with impervious surfaces that absorb, retain, and reradiate more solar energy than do grass and trees. The rate of this effect depends on the physical properties of different surface types, their configuration within the urban fabric, regional meteorology, and localized microclimate. As average air temperatures rise, so does the urban heat island effect. The annual mean air temperature of a city with at least 1 million people can be more than 5°F warmer than surrounding rural areas. Some project that daytime temperatures in US cities will increase by up to 10°F on average by the end of the 21st century. The urban heat island effect creates increased demand for electricity for cooling and consequently increased power-plant-related GHG emissions.

---

[66]   Score Card, *supra* note 23, at 42, 64.
[67]   Score Card, *supra* note 23, at 64.
[68]   Score Card, *supra* note 23, at 49–56.
[69]   Score Card, *supra* note 23, at 56.

Cities are tackling the problem in a variety of ways; in the forefront are land development policies that increase and preserve vegetated land. At the same time, they are requiring and/or offering incentives for the installation of cool roofs. New pavements are using highly reflective coatings to reflect away rather than absorb heat. Tree canopies are being built.[70] Milwaukee, Wisconsin enacted a 200,000 street trees program.[71] Urban forests are being grown. They are a collection of trees and other vegetation found within the built urban environment; they are primarily human-created—the result of tree planting and greening activities carried out by people rather than a native forest ecosystem. Urban forests serve to sequester atmospheric CO_2 and are long-term carbo sinks. In the San Francisco urban forests,[72] store more than 196,000 tons of carbon and more than 260 tons of atmospheric pollutants are filtered out of the air annually.[73] Other cities with urban forests programs include Philadelphia,[74] Columbus, Ohio,[75] Orlando, Florida,[76] and New York.[77]

---

[70] America's Pledge, *supra* note 16, at 50. More than 3,400 communities have committed to implementing basic urban forestry standards through Tree City, USA, including maintaining a tree board or department, and with tree ordinances. *Id.*

[71] i-Tree Ecosystem Analysis – Milwaukee, Urban Forest Effects and Values, September 2008.

[72] Based on an estimate of, on average, 774 gallons intercepted annually per tree (Davey Resource Group 2013).

[73] *Assessing Urban Forest Effects and Values: San Francisco's Urban Forest*, United States Forest Service (2007). In 2016, San Francisco voters approved Proposition E to amend the City Charter to transfer responsibility for the care of the City's 124,000-plus trees and surrounding sidewalks from property owners to Public Works. The ballot measure took effect July 1, 2017. www.sfgov.org/sfplanningarchive/urban-forest-plan.

[74] www.TreePhilly.org.

[75] www.columbus.gov/recreationandparks/urban-forestry.

[76] www.orlando.gov/initiatives/2018-Community-Action-Plan.

[77] New York City's Urban Forest, https://www.nycgovparks.org/trees; *see also* Elmqvist, T., Setälä, H., Handel, S. N., Van Der Ploeg, S., Aronson, J., Blignaut, J. N., ... & De Groot, R. (2015). Benefits of restoring ecosystem services in urban areas. Current opinion in environmental sustainability, 14, 101–108. doi:10.1016/j.cosust.2015.05.001; Elmqvist, T., Setälä, H., Handel, S. N., Van Der Ploeg, S., Aronson, J., Blignaut, J. N., ... & De Groot, R. (2015). Benefits of restoring ecosystem services in urban areas. Current opinion in environmental sustainability, 14, 101–108. doi:10.1016/j.cosust.2015.05.001; Driscoll A.N., Ries, P.D., Tilt, H.J., Ganio, L.M. (2015). Needs and barriers to expanding urban forestry programs: An assessment of community officials and program managers in the Portland-Vancouver metropolitan region. 14(1):48–55. doi.org/10.1016/j.ufug.2014.11.004.

I Financial or Nonfinancial Incentives

Many states and cities offer financial incentives to owners to employ carbon energy technologies and to build renewable energy projects. There are tax abatements, permit fee reductions or waivers, grants, and rebates. Some also have policies that provide financing and loans for efficiency upgrades and solar installation.[78] Nonfinancial incentives, such as accelerated permitting and density bonuses, are also used to encourage developers and builders to construct buildings that exceed code minimums and meet additional certifications like LEED.[79]

The energy financing platform that is growing in stature is the green bank.[80] Green banks can be public, quasi-public, or non-profit entities that mobilize private capital for investments in clean energy and energy efficiency. The New York City Energy Efficiency Corporation (NYCEEC) was created in 2010 and, now just past its tenth year, it boasts of many different kinds of impact—economic (capital mobilized), environmental (CO_2 eliminated), and social (affordable and efficient housing created).[81]

IV ADAPTATION THROUGH NATURAL SYSTEMS: ECOLOGICAL DESIGN

In a less urgent climate, the natural systems worked ably and kept the sea waters at bay. Mangroves and salt marshes captured moving sediments to help reduce waves; reefs acted as breakwaters; and rock and sand emerged as islands and beaches. Alas, human activity—development pressures, profligate energy uses—have overwhelmed these ordinary climate forces, prompting the need for more purposeful human efforts to support our natural systems. As it pertains to our seashores' threats, some governments are taking actions to fortify these natural mechanisms for protecting seashores. In Connecticut, the Coastal Resilience Program[82] was the first in the nation to have assessed the

---

[78] Property Assessed Clean Energy: PACE Nation, "PACE Programs," http://pacenation.us/pace-programs/.

[79] America's Pledge, *supra* note 16, at 42.

[80] NREL Transforming Energy, State, Local and Tribal Governments, https://www.nrel.gov/state-local-tribal/basics-green-banks.html.

[81] *See* Erin Muir and Satyajit Bose, *The Green Bank Opportunity: Mobilizing Capital for Low-Carbon Energy in Buildings, Financing Innovations from New York City Energy Efficiency Corporation*, The HSBC Centre of Sustainable Finance, https://www.sustainablefinance.hsbc.com/mobilising-finance/the-green-bank-opportunity.

[82] *See* Coastal Resilience, Connecticut, https://coastalresilience.org/project/connecticut/#:~:text=The%20Coastal%20Resilience%20Program%20provides,across%20the%20state%20of%20Connecticut.

entire coastline for future salt marsh advancement zones (where salt marshes will likely advance with projected sea-level rise) down to the parcel scale.[83] In Maryland, a living shorelines initiative involves planting native wetland plants, grasses, shrubs, and trees at various parts along the tidal water line.[84] The Chesapeake Bay is the largest and most productive estuary in the United States.[85] In collaboration with New York City communities, New York State has seeded the New York Harbor with oysters that diminish waves and harness shellfishes' biotic processes to clean millions of gallons of harbor water.[86]

But what about development elsewhere, away from the shores? What can be done to nourish and support nature's powers to protect? Ecological design is a new strategy, practiced by planners and architects, that promises resiliency and sustainability.[87] Ecological design embraces an alternative vision to create the space for and set into motion ecological processes that mimic Mother Nature in an intentional and measured way.[88] It requires consideration for the environmental impacts of development for the whole lifecycle. Sim Van der Ryn and Stuart Cowan coined the term, meaning "any form of design that minimizes environmentally destructive impacts by integrating itself with living processes."[89] It embraces architecture efforts, sustainable agriculture, ecological engineering, ecological restoration, and other fields. The broad aim is to cause us to live and build with a conscious sense of stewardship, to maintain our connections to the natural world, and to elevate nature to the other imperatives of living, rather than as an afterthought. In ecological design, water is safe to consume as natural systems avert or capture pollutants. Food supply is enhanced as soils are productive, and habitats for wildlife and fish

---

[83]   *See* Coastal Resilience, Connecticut First for Future Saltmarsh Advancement Assessment at Fine Scale for Entire Coast, https://coastalresilience.org/connecticut -first-for-future-saltmarsh-advancement-assessment-at-fine-scale-for-entire-coast/.

[84]   *See* Maryland Department of Natural Resources, Living Shorelines, https://dnr .maryland.gov/ccs/Pages/livingshorelines.aspx.

[85]   National Ocean Service, *Where is the largest estuary in the United States?*, https:// oceanservice.noaa.gov/facts/chesapeake.html#:~:text=District%20of%20Columbia .-,The%20Chesapeake%20Bay%20is%20the%20largest%20estuary%20in%20the %20United,and%20the%20District%20of%20Columbia.

[86]   The Billion Oyster Project, https://www.billionoysterproject.org/; *see generally*, North Atlantic Coast Comprehensive Study: Resilient Adaptation to Increasing Risk, https://www.nad.usace.army.mil/Portals/40/docs/NACCS/NACCS_main_report.pdf.

[87]   George F. Thompson, Frederick R. Steiner, Armando Carbonell, *Nature and Cities: The Ecological Imperative in Urban Design and Planning*, Landlines, Winter 2016.

[88]   Sᴉᴍ Vᴀɴ Dᴇʀ Rʏɴ & Sᴛᴜᴀʀᴛ Cᴏᴡᴀɴ, Eᴄᴏʟᴏɢɪᴄᴀʟ Dᴇsɪɢɴ 3–4, 47 (10th ed. 2007).

[89]   Vᴀɴ ᴅᴇʀ Rʏɴ S, Cᴏᴡᴀɴ S., Eᴄᴏʟᴏɢɪᴄᴀʟ Dᴇsɪɢɴ (1996) 18; *see also* Martin Charter (2019). "Designing for the Circular Economy". Abingdon, p.21.

are nurtured. Flooding is reduced as rainwater is captured in rain gardens or by grass and returned to its natural course. Reliance on fossil fuels for energy is reduced as renewable sources, green materials[90] are used, and off-grid energy systems are established.

V   ADAPTING THE BUILT ENVIRONMENT

States' and cities' efforts to anticipate a rapidly changing climate are wide-ranging and aggressive—the programs and systems range from planning and zoning to smart homes to urban forests to district heating.[91] As the list of measures is almost endless, this chapter strives only to spotlight some of the more prominent measures, with a few examples.

A   Planning

The first step in developing adaptation strategies is to identify the vulnerabilities of the system or sector interest and the climate risks to that system. Those vulnerabilities should measure socio-economic and environmental conditions and the impacts of climate change. From there, adaptation measures are considered vis-à-vis the risk, keeping in mind the expected effectiveness, cost, and feasibility of the measures selected.

In the last decade or so, states and many more local governments have begun to assess anticipated climate change impacts and craft adaptation strategies.[92] Nearly half of the states have completed, or are in the process of completing,

---

[90]   EcoMaterials, such as the use of local raw materials, are less costly and reduce the environmental costs of shipping, fuel consumption, and CO2 emissions generated from transportation.

[91]   The district heating and cooling system (DHCS) of the energy company Helen Oy in Helsinki represents state-of-the-art technology that contributes to both climate change mitigation and adaptation. It has significantly increased the energy efficiency of buildings, improving air quality in Helsinki, and simultaneously providing an energy-efficient adaptation tool to avoid conventional air-conditioning in the summer. District heating and cooling can also be a major resilience investment by reducing the risk and impact of power outages. Although the DHC system in Helsinki is still partially based on fossil fuels, energy savings, using combined heat and power production, are equivalent to the consumption of 500,000 detached homes with conventional systems.

[92]   *See* Vicki Arroyo & Terri Cruce, *State and Local Adaptation, in The Law Of Adaptation to Climate Change*, at 569–87; *see also State and Local Adaptation Plans*, Georgetown Climate Ctr., http://www.georgetownclimate.org/node/3325; ICLEI-Local Governments For Sustainability USA, Local Governments, Extreme Weather, And Climate Change 2012 (2012), *available at* http://www.icleiusa.org/action-center/learn -from-others/local-governments-extreme-weather-and-climate-change-2012 (detailing actions taken by 20 US cities).

climate adaptation plans as comprehensive guides, setting out an overall policy to specifically address the impacts of climate change, goals, and milestones.[93] The California Climate Action Plan[94] is a comprehensive plan addressing all impacts of climate change, backed up by a recent amendment to the Public Resources Code that requires agencies to take into account climate change when planning, designing, and investing in infrastructure.[95] New Hampshire's Comprehensive Master Plan 123 requires zoning ordinances that ensure walkability, green infrastructure, sustainable building, and permeable pavements.[96] The Pittsburgh Comprehensive Plan calls for 22,000 acres of publicly owned land to be transformed into green infrastructure, community gardens, urban farms, greenways, and parklets.[97] Following Superstorm Sandy in 2012, New York State and the City of New York adopted a host of plans to anticipate the next superstorm, including downscale zoning that responds to local climate conditions,[98] and measures to reduce flooding by stabilizing the coastal edge, discouraging the development of at-risk locations, and mitigating negative impacts of new projects; improving stormwater and wastewater management;

---

[93] *See* Arroyo & Cruce, *supra* note 92, at 574.

[94] California Climate Action Plan, www.climatechange.ca.gov.

[95] For example, Section 71155 is added to the Public Resources Code, to read: 71155. (a) Consistent with this part, state agencies shall take into account the current and future impacts of climate change when planning, designing, building, operating, maintaining, and investing in state infrastructure. (b) (1) By July 1, 2017, the agency shall establish a Climate-Safe Infrastructure Working Group for the purpose of examining how to integrate scientific data concerning projected climate change impacts into state infrastructure engineering, including oversight, investment, design, and construction.

[96] Adopted September 2010. *See also* Chicago Climate Task Force, *Chicago Climate Action Plan* (2008), http://www.chicagoclimateaction.org/; *City of Portland and Multonah County, Climate Action Plan*, 2009, Year Two, Progress Report (April 2012); San Diego Climate Action Plan, https://www.sandiego.gov/planning/genplan/cap; PlaNYC, http://www.nyc.gov/html/planyc/html/resiliency/resiliency.shtml (multifaceted guide, covering housing, transport, solid waste, heating oils); *Preparing For Climate Change* (2007 King County Climate Plan), Wolf, K., 2009: Adapting to Climate Change: Strategies from King County, Washington. *PAS Memo*, March/April, 11, http://www.planning.org/pas/memo/previous.htm; http://www.kingcounty.gov/depts/dnrp/wlr/sections-programs/river-floodplain-section/documents/flood-hazard -management-plan-update.aspx. *see generally*, U.S. Conference of Mayors, Taking Local Action: Mayors and Climate Protection Best Practices 5 (2014), http://usmayors .org/climateprotection/2014awardees/0620-report-climateprotectionbp.pdf.

[97] Roy Kraynyk, Allegheny Land Trust, *Using Transfer Development Rights to Facilitate and Sustain Community Green Space and Gardens*, (2017) https://alleghenylandtrust.org/wp-content/uploads/2017/09/20170919_TDRWhitepaperv2.0 .pdf.

[98] http://stormrecovery.ny.gov/sites/default/files/documents/CRZ_Guidance.pdf.

making power supply more resilient and redundant; enhancing emergency preparedness and response; shoring up commercial corridors and critical supply
chains; and improving residential resiliency.[99]

Cites are also acting to remove institutional and legal barriers that historically have stood in the way of implementing adaptation and resiliency measures, such as limits on mixed-uses within zones, parking space requirements,
and siting for solar panels.[100] There are scores of guides, toolkits, and reports
by nonprofit think tanks and research institutions.[101]

Despite seeming aggressive, states and local governments that have taken
measures are in the minority. Many states have not taken substantial steps to
address climate adaptation; the number of states without adaptation plans is
twice that of those that have them. Indeed, some state legislators have barred
initiatives toward the end of climate adaptation.[102]

---

[99]   The Community Risk and Resiliency Act (CRRA), Ch. 355 Laws 2014, amend-
ing the Environmental Conservation Law and Public Health Law included five major
provisions and required applicants for permits or funding in a number of specified programs to demonstrate that future physical climate risk due to sea-level rise, storm surge,
and flooding had been considered in project design, and that DEC consider incorporating these factors into certain facility-siting regulations. The CRRA added mitigation of
risk due to sea-level rise, storm surge, and flooding to the list of smart-growth criteria to
be considered by state public-infrastructure agencies. Guidance on Natural Resilience
Measures – The CRRA required DEC, in consultation with the Department of State
(DOS), to develop guidance on the use of natural resources and natural processes to
enhance community resilience. Model Local Laws Concerning Climate Risk – CRRA
required DOS, in cooperation with DEC, to develop model local laws to increase
community resilience. The 2019 Climate Leadership and Community Protection Act
(CLCPA, https://climate.ny.gov/) amended the CRRA to expand the scope of the
CRRA to require consideration of all climate hazards, not only sea-level rise, storm
surge, and flooding, in these permit programs.

[100]   *See* Sara C. Bronin, *Rezoning the Post-Industrial City: Hartford*, 31 Prob. &
Prop. 44 (2017) (describing the process of overhauling the zoning code for efficiency
and sustainability).

[101]   *See e.g.*, Sustainable Communities Development Code: A Code for the 21st
Century, by Rocky Mountain Land Use Institute, https://www.dvrpc.org/announce/
2008-10_GreenCodes/SustainableCommunityDevelopmentCodeBetaVersion1.1.pdf;
International Council for Local Environmental Initiatives [hereinafter ICLEI], https://
iclei.org/.

[102]   In 2012, North Carolina adopted legislation that ignored state scientists' sea-level
rise predictions and barred further statewide predictions for four years, chilling the
state's role in facilitating local planning for sea-level rise. *See* Wade Rawlins, *North
Carolina Lawmakers Reject Sea Level Rise Predictions*, Reuters (July 3, 2012), http://
www.reuters.com/article/2012/07/03/us-usa-northcarolina-idUSBRE86217I20120703
(describing initial legislation, which required that projections be based on historic
trends); Bruce Henderson, *North Carolina's Coast is 'Hot Spot' for Rising Sea Levels*,
Charlotte Observer (Nov. 5, 2012), http://www.mcclatchydc.com/2012/11/05/173575/

B   Structural Design and Building Codes

Nationwide, building codes are being updated to require fortifications—some adopting FEMA's standards, others developing their own.[103] In some states, insurance incentives are offered for fortifying structures.[104] Fortification of buildings includes elevations (of sites, structures, and critical systems), the use of wind- and water-resistant materials,[105] fire-safe design, and emergency back-ups.[106] Some cities are demolishing rickety buildings.[107] Other cities

---

north-carolinas-coast-is-hot-spot.html (describing final legislation, which delayed but did not prevent climate change-influenced sea-level rise projections).

[103] "On May 4, 2007, an EF-5 tornado struck the City of Greensburg, Kansas, destroying more than 90% of its building stock. In the wake of the disaster, the community set forth to rebuild and become a model sustainable rural community. The city adopted a Long-Term Community Recovery Plan22 in 2007, prepared through FEMA's Long-Term Community Recovery (LTCR) program." The program led to a sustainable, comprehensive plan as the blueprint for all new development and for rebuilding. The "Greensburg Sustainable Comprehensive Master Plan23" contains an entire section dedicated to "hazard mitigation, focusing on tornado, thunderstorm, and other high windstorm hazards." It calls for the "integrati[on] of hazard mitigation into the recovery plan or land development code by requiring that power lines be buried to reduce damage and decrease the frequency of power outages." It also "require[s] back-up generators for critical facilities and test them regularly." Another measure that calls for the use of native species in the local land development code or tree ordinance and the use of native plants and trees for ornamental plantings to decrease vegetation damage and as a brace against winds. Building codes would be strengthened to reduce wind-related damages. Safe rooms in accordance with FEMA guidelines would be built. FEMA Hazard Mitigation Guide, supra note 59, at § 5–9.

[104] *See e.g.*, Ala. Code¶ 56,110. Cost of Buildings, Improvements, Restoration; Green, *supra* note 14, at 551.

[105] Colorado Springs bans wood shake roofs and requires roofing materials and assembly to keep fire from penetrating the roof and igniting the structure below. From 2002 to 2016, more than 69,000 roofs were replaced or upgraded to fire-resistant roofing. City of Colorado Springs Fire Department, *Ignition Resistant Construction Design Manual*, (Colorado Springs, CO: City of Colorado Springs Fire Department, 2016), https://coloradosprings.gov/sites/default/files/final_hillside_wildfire_ mitigation_design_manual_final_document_third_printing.pdf; *see generally* Urb. Green Council, *Building Resiliency Task Force*, 14 (2013), http://issuu.com/urbangreen/docs/brtf_executive_summary.

[106] PlaNYC, *supra* note 96, at 126, 129 (hookups for access to generators, anti-backflows, and faucets in common areas).

[107] Kellen Zale, *Urban Resiliency and Destruction*, 50 Idaho L. Rev. 85, 86 (2014) (discussing destroying buildings to create resiliency, Zale asserts that "...destruction is as necessary to urban resiliency as creation. Destruction allows cities to eliminate outdated, underutilized, and vacant buildings; create the necessary physical space for redevelopment and innovation; and redirect the city's economic resources to best meet the needs of residents." Following the second 1,000-year flood in two years (2016 and

have adopted complex permitting requirements that require a developer to demonstrate how the project will respond to higher temperatures, heatwaves, potential flooding, and storm intensity/actions. The proposal must include plans to mitigate those impacts (such as by reducing energy consumption and demand, preventing interruption, reducing urban heat island effects, accommodating more rainfall and extreme weather, resisting flooding, wind, and wave impacts) and provide for the elevation of structures or critical systems.[108] Some cities are prescribing limits on landscaping, requiring more open space, recommending plants with erosion control qualities, and low maintenance.[109]

C   Protection from the Rising Seas

On the coasts, cities are requiring buffers (walls and dunes) and setbacks from seashores.[110] In Maine, the Sand Dune Rules require that structures greater than 2,500 square feet be set back at a distance calculated based on the future shoreline position and considering two feet of sea-level rise over the next 100 years.[111] Rhode Island requires public agencies considering land-use

---

2018), historic Ellicott City, Maryland, proposed, among other things, razing at least four historic buildings that would leave between a maximum water depth of 4 feet and 5 feet on the lower main street, in the event of an event similar to the 2016 flood. https://www.baltimoresun.com/maryland/howard/ph-ho-cf-flood-announcement-0510 -story.html.

[108] For   Boston,   *see*   https://www.cityofboston.gov/Images_Documents/A %20Climate%20of%20Progress%20-%20CAP%20Update%202011_tcm3-25020.pdf.

[109] *Id.*,   https://www.cityofboston.gov/Images_Documents/A%20Climate%20of %20Progress%20-%20CAP%20Update%202011_tcm3-25020.pdf at 22-23. *See also* The City of Denver, https://www.denvergov.org/content/dam/denvergov/Portals/747/ documents/Natural_Areas/DenverPark_LandscapeTypologyManual.pdf.

[110] Scientists now believe that hard armoring can actually exacerbate coastal erosion and beach loss in the face of rising seas, diminishing both the protective function of natural shorelines and the beaches we treasure. In addition, by bouncing waves back into the ocean, seawalls can harm local wildlife, erode beaches, and increase the impacts of storms. In any case, hard armoring usually does not protect against the infiltration of saltwater from below, which may have deteriorating effects on historic structures. Urban Green, Building Resiliency Task Force 14 (2013), *available at* http://issuu .com/urbangreen/docs/brtf_executive_summary (last visited Jan. 23, 2015). *See generally,* Union of Concerned Scientists, *Encroaching Tides, How Sea Level Rise and Tidal Flooding Threaten U.S. East and Gulf Coast Communities over the Next 30 Years,* Encroaching-tides-full-report.pdf. at 42. www.usa.org. Encroaching Tides, at 7, 13, 14 (steep slope mountain ridge protection; maximum grading allowances; preservation of green space).

[111] http://s3.amazonaws.com/nca2014/low/NCA3_Climate_Change_Impacts_in _the_United%20States_LowRes.pdf?download=1.

applications to accommodate a 3- to 5-foot rise in sea level.[112] California has adopted a Sea Level Rise Guidance, which contains probabilistic sea-level rise projections and worst-case scenarios, integrating both with an adaptive pathways approach that encourages robust and flexible plans that can adjust if seas rise faster than expected.[113] Boston's Climate Resilient Design Standards and Guidelines include designs for various barriers: vegetated berms, seawalls, raised roadways, and deployable barriers.[114] After four hurricanes made landfall in 2004, and after Superstorm Sandy, the city of Cedar Key, Florida, adopted a plan that includes a living shoreline program.[115] Some cities are encouraging and facilitating, through loans, the elevation of homes in flood-prone areas.[116]

Where the barriers and setbacks might prove ineffective, cities are planning managed retreat programs. Managed retreat means the voluntary movement and transition of people and ecosystems away from vulnerable coastal areas. A gradual, proactive relocation strategy is seen as more efficacious than the typical reactive responses to sea-level rise and coastal storms after the harm has been experienced.[117] To help facilitate a managed retreat program, the

---

[112] *Id.*

[113] https://www.opc.ca.gov/2013/04/update-to-the-sea-level-rise-guidance-document/.

[114] https://www.boston.gov/sites/default/files/embed/file/2018-10/climate_resilient_design_standards_and_guidelines_for_protection_of_public_rights-of-way_no_appendices.pdf at 10-13, 19.

[115] Living shorelines: enhancing natural habitat along the shoreline to dampen wave energy and accumulate sediment. The benefits are cost-effective, likely to adapt to changing sea levels, benefits environmental function (e.g., habitat, water quality), ability to gain (accrete) land by trapping sediments, less likely to fail and require repairs (high resilience), likely to be long-lasting. There are drawbacks: they may not be suitable for all sites, can be challenging to implement (because living shorelines are a relatively new approach, they lack streamlined permitting mechanisms and technical expertise is not widespread—but this situation is improving). https://ufl.maps.arcgis.com/apps/MapJournal/resources/tpl/viewer/print/print.html?appid=f8473d6f257b4e6bab4d42c343e3901d.

[116] http://shoreupct.org/. *See also* https://www.miamidade.gov/environment/repetitive-losses.asp – see "Paying for Flood Mitigation;" https://www.miamibeachfl.gov/wp-content/uploads/2017/08/Rising-Above-the-Risk_FAQs-05052017.pdf; https://www.charleston-sc.gov/2386/Flood-Mitigation-Resources.

[117] The Alaskan community of Shishmaref elected to relocate in full from its current site onto the mainland 5 miles away. Global warming is happening twice as fast in Alaska than anywhere else on earth. Loss of ice meant loss of buffer against severe storm surges. This, combined with the effects of thawing permafrost, meant the softening of the very land the village was built upon, a loss of 3–5 feet of shoreline per year, with a single severe storm washing away 50 feet of land. In 1997 and 2002, some homes actually fell into the ocean, and several more needed to be moved. https://www.nytimes.com/2016/08/20/us/shishmaref-alaska-elocate-vote-climate-change.html.

Georgetown Climate Center's Managed Retreat Toolkit offers a range of legal and policy tools for state and local governments, including advice on planning, building, and altering infrastructure.[118]

D   Green Infrastructure

Green infrastructure means using plant or soil systems, permeable surfaces or substrates, stormwater harvest, and reuse, or landscaping to store, infiltrate, or evapotranspirate stormwater.[119] It is a cost-effective, resilient approach to managing wet weather impacts that also provides many community benefits.[120] Many communities are revising their land development standards to require the incorporation of green infrastructure.[121] The Louisville, Kentucky Green

---

[118] Georgetown Management Retreat Toolkit. https://www.georgetownclimate.org/adaptation/toolkits/managed-retreat-toolkit/introduction.html.

[119] What is Green Infrastructure, https://www.epa.gov/green-infrastructure/what-green-infrastructure.

[120] *See generally* Josh Foster, Ashley Lowe, Steve Winkelman, *The Value of Green Infrastructure for Urban Climate Adaptation*, The Center for Clean Air Policy, February 2011, www.ccap.org_assets_The-Value-of-Green-Infrastructure-for-Urban-Climate-Adaption_ccap-Feb-2011.pdf. (describing the principles and efficacy of green infrastructure measures).

[121] Josh Foster, Ashley Lowe, Steve Winkelman, *The Value of Green Infrastructure for Urban Climate Adaptation*, The Center for Clean Air Policy, February 2011, www.ccap.org_assets_The-Value-of-Green-Infrastructure-for-Urban-Climate-Adaption_ccap-Feb-2011.pdf; Appendix 1–6. To meet the city's ambitious green infrastructure goals, Philadelphia's Public Works Department (PWD) developed a three-pronged strategy: 1) invest in greening public property and rights-of-way, integrating green infrastructure into public space improvements, including street, sidewalk, and park projects; 2) require green infrastructure investments for new development and redevelopment on private property; permit regulations require new development and redevelopment projects that disturb more than 15,000 square feet of land install/maintain green infrastructure sufficient to manage the first inch of stormwater that falls on the site; and 3) encourage voluntary retrofits by existing private parcel owners. The Greened Acre Retrofit Program incentivizes "private parcel retrofits by modifying commercial property owners' monthly stormwater fees to reflect each property's relative contribution to stormwater runoff" by assessing stormwater fees based on the size of impervious areas on individual lots. There are incentives to encourage property owners to install green infrastructure practices sufficient to manage the first inch of stormwater runoff—a savings of up to 80% on their monthly stormwater fees. Also, the City of Portland, Oregon has adopted a comprehensive green infrastructure program, using bioswales and rain gardens, among other things. *City of Portland and Multonah County, Climate Action Plan*, 2009, Year Two, Progress Report (April 2012), *see also* International Council for Local Environmental Initiatives & World Wildlife Fund, *Measuring Up 2015: How U.S. Cities are Accelerating Progress Toward National Climate Goals* 35, www.worldwildlife.org/climate.

Infrastructure Program builds and promotes green infrastructure, aiming to reduce stormwater runoff and incentivize opportunities for green installations.[122] The Wilmington, Massachusetts Demonstration Project promotes permeable paving materials and bioretention basins in parking lots to reduce the quantity of stormwater runoff and non-point source pollution.[123] The city of Mesa, Arizona, has adopted the Low-Impact Development ToolKit. The kit speaks about green streets with curb cuts to direct water into vegetated areas. This measure will also enhance landscaping, restore a natural wash, or create wetlands, helping trees to grow faster and healthier. The toolkit also offers guidance on installing vegetated swales, vegetated retention basins, cells, and planters.[124]

With green infrastructure, natural wetlands can be used for the infiltration of wastewater, onsite vegetated swales, as opposed to curbs, capture and redirect rain and stormwater. Low-water use plants—xeriscaping—reduces water consumption.[125] Bio-retention basins and vegetated swales[126] capture water from runoff. Vegetated roofs help to control the urban heat island effect and minimize water runoff.[127] Downspout disconnection programs aim to prevent clear water from entering and overwhelming the city's wastewater treatment plant, and instead allow it to drain into the ground.[128]

Rain gardens, which are bowl-shaped areas, capture and filter stormwater runoff from roofs, driveways, and other hard surfaces (indeed, more than

---

[122] https://louisvillemsd.org/Green.

[123] https://www.mass.gov/service-details/demonstration-3-permeable-paving
-materials-and-bioretention-in-a-parking-lot.

[124] City of Mesa, Arizona, *Low Impact Development ToolKit*, https://www.mesaaz
.gov/home/showdocument?id=14999#:~:text=The%20Toolkit%20is%20intended
%20to,more%20sustainable%20stormwater%20design%20practices.

[125] "Xeriscaping is a systematic method of promoting water conservation in land-scaped areas. Although xeriscaping is mostly used in arid regions, its principles can be used in any region to help conserve water. There are seven basic xeriscaping principles: planning and design, selecting and zoning plants appropriately, limiting turf areas, improving the soil, irrigating efficiently, using mulches, and maintaining the landscape. United States Department of Energy, *Landscaping Water Conservation*, https://www.energy.gov/energysaver/design/landscaping-energy-efficient-homes/landscaping-water-conservation; *see also* EPA, *Water-Smart Landscapes*, https://www.epa.gov/sites/production/files/2017-01/documents/ws-outdoor-water-efficient-landscaping.pdf (detailing tips for water-smart landscaping, a concept similar to xeriscaping, and how to apply this concept, such as by choosing plants native to the location that need only natural rainwater).

[126] https://dnr.wi.gov/topic/Stormwater/standards/index.html.

[127] https://louisvillemsd.org/Green.

[128] https://www.mmsd.com/what-you-can-do/downspout-disconnection. *See also* Portland, https://www.portlandoregon.gov/bes/54651.

30% more water than a regular lawn). The water is filtered into the ground, keeping it from becoming harmful water pollution laden with household fertilizers, pesticides, oils, and other contaminants coursing from our roofs, lawns, driveways, or parking lots.[129] They are also nurturing places for birds and butterflies.[130] The Washington State University and Stewardship Partners are pioneering a campaign to install 12,000 rain gardens in the Seattle/Puget Sound Region.[131] The campaign goal would absorb up to 160 million gallons of polluted runoff to protect waterways.[132]

Rain harvesting involves capturing and storing rainwater for various domestic uses. New York City has launched a Rain Barrel Giveaway Program. In the program, rain barrels are connected directly to the property's gutter or downspout to capture and store rainwater that falls on your rooftop. The stored rainwater can be used for outdoor chores, like gardening or washing your car, which ordinarily can account for up to 40% of an average household's water use during the summer. The rain barrels also help reduce the number of combined sewer overflows that enter the city's sewer system, which protects the health of local waterways. The barrels are free.[133] In 2016, the State of Colorado amended its state constitution to allow homeowners to collect rainwater from a rooftop if no more than two rain barrels, with a combined storage capacity of 110 gallons or less, are utilized; the property is used primarily as a single-family residence or a multi-family residence with four or fewer units; the water is used for outdoor purposes including irrigation of lawns and gardens and not for drinking or indoor household purposes; and the water is used on the residential property on which the precipitation is collected.[134]

Alleys and streets are being greened by planting grass and trees where pavement and concrete once existed. Edmonston, Maryland's "the Greening

---

[129] Rain gardens are relatively simple to install and feature well-draining soil and easy-to-maintain plants that allow for stormwater infiltration. Some may include bioretention facilities, stormwater planters, and bioswales. http://www.soundimpacts.org/projects/list/type/rain-garden; https://dnr.wi.gov/topic/Stormwater/raingarden/.

[130] *Id.*

[131] https://www.12000raingardens.org/.

[132] *Id.*

[133] https://www1.nyc.gov/site/dep/whats-new/rain-barrel-giveaway-program.page.

[134] Texas and Ohio allow rainwater harvesting for potable purposes, a practice that is frequently excluded from other states' laws and regulations. Rhode Island, Texas, and Virginia offer tax credits or exemptions on the purchase of rainwater harvesting equipment. Oklahoma passed the Water for 2060 Act in 2012 to promote pilot projects for rainwater and graywater use, among other water conservation approaches. State Rainwater Harvesting Laws and Legislation, https://www.ncsl.org/research/environment-and-natural-resources/rainwater-harvesting.aspx.

of Decatur Street Project"[135] included the building of native tree canopies, permeable-surfaced bike lanes, and tree boxes. The project promised to capture stormwater pollution before it enters the Anacostia River, reduce flooding along Decatur Street, and slow traffic to create a safe and pleasant environment for pedestrians and cyclists.[136] Chicago has promulgated a Green Alley Handbook, a guide for creating a "greener and environmentally sustainable Chicago."[137]

Tree canopies make for beautiful street vistas, but they also serve more organic purposes, including preventing soil erosion from heavy precipitation, providing shade for cooling, and absorbing much CO_2. In the San Diego Climate Action Plan, the city has undertaken to cover 35% of the city with tree canopies by 2035.[138]

While green infrastructure serves to add beauty to the environment, protect wildlife habitat, conserve water, and limit pollution of water sources, perhaps its most essential function is stormwater management. When rain falls in natural, undeveloped areas, the water is absorbed and filtered by soil and plants. But, when rain falls on our roofs, streets, and parking lots, the water cannot soak into the ground as it should. Stormwater drains through gutters and other engineered collection systems, carrying trash, bacteria, heavy metals, and other pollutants from the urban landscape. Much of this runoff ends up in our sewer systems. The sewer systems in many older American cities were designed to handle both sewer and stormwater runoffs. But, these ancient systems lack the capacity to handle the increased loads from the rising populations and increasing major storms. New York City, which has a massive system with 14 treatment plants serving millions of customers, faces annual

---

[135] The city had suffered severe flooding four years in a row due to development, a poor stormwater management system, and a high percentage of impervious cover (e.g., roadways, rooftops, parking lots). A flood in 2006 left 56 homes in the town partially underwater, and most people lost everything they owned. Residents decided it was time for a change and envisioned a safer community that could also be a leader in environmental stewardship. https://www.epa.gov/sites/production/files/2016-05/documents/edmonston.pdf.

[136] *Id.*

[137] https://www.chicago.gov/content/dam/city/depts/cdot/Green_Alley_Handbook_2010.pdf.

[138] San Diego plans to cover 35% of the city with tree canopies by 2035. https://www.sandiego.gov/sustainability/climate-action-plan (at 41). New York City launched a million trees program in 2007. https://www.nycgovparks.org/trees/milliontreesnyc. *See also* Keene, New Hampshire's Comprehensive Master Plan 123 (September 2010) (requiring zoning ordinance that ensures walkability, green infrastructure, sustainable building, and permeable pavements).

overflows of roughly 30 Bgal./year.[139] Kansas City, Missouri's wastewater system serves more than 650,000 customers and faces an average annual overflow volume of 6.4 Bgal./year. The city is proposing an overflow control plan, which aims to capture 88% of these flows. It is estimated to cost $2.5 billion by 2035, constituting the largest infrastructure project in the city's history.[140] In addition to major and costly sewer system rebuild, cities are employing green stormwater infrastructure (GSI). GSI involves the use of various technological and natural devices to divert and capture water during rainfall events, either temporarily (with eventual flow back to the system) or permanently (through infiltration or evapotranspiration). GSI uses vegetation, soil, or other natural elements[141] by naturally filtering stormwater runoff and improving water

---

[139] New York City Department of Environmental Protection, undated; Jordan R. Fischbach, Kyle Siler-Evans, Devin Tierney, Michael T. Wilson, Lauren M. Cook, and Linnea Warren May, *Robust Stormwater Management in the Pittsburgh Region: A Pilot Study*, Santa Monica, Calif.: RAND Corporation, RR-1673-MCF at 13, 2017.

[140] Kansas City, Missouri, Water Services Department, 2012.

[141] *Id.* at 18. GSI as a component of future long-term sewer overflow control plans (USEPA, 2014), and the approach is being studied and implemented at scale as part of LTCPs or other climate adaptation efforts in other major metropolitan areas, including Philadelphia; Seattle; Washington, D.C.; and New York City of Environmental Protection, 2010; Philadelphia Water Department, 2009, District of Columbia Water and Sewer Authority, 2015, District of Columbia Water and Sewer Authority, "Long Term Control Plan Modification for Green Infrastructure," May 2015. As of March 20, 2017: https://www.dcwater.com/sites/default/files/green -infrastructure-ltcp-modificaitons.pdf; New York City Department; Kansas City, Missouri, Water Service Department, *Overflow Control Program: Overflow Control Plan*, January 2009, last updated April 30, 2012. As of March 20, 2017: https:// www.kcwaterservices.org/wp-content/uploads/2013/04/; Overflow_Control_Plan_ Apri302012_FINAL.pdf; Milwaukee Metropolitan Sewerage District, *Regional Green Infrastructure Plan*, Milwaukee, Wisc., June 2013, http://www.freshcoast740.com/ -/media/FreshCoast740/Documents/GI%20Plan/Plan%20docs/MMSDGIP_Final.pdf; *NYC Green Infrastructure Plan: A Sustainable Strategy for Clean Waterways*, New York: City of New York, Office of the Mayor, 2010, http://www.nyc.gov/html/ dep/pdf/green_infrastructure/NYCGreenInfrastructurePlan_LowRes.pdf; Philadelphia Water Department, *Philadelphia Combined Sewer Overflow Long Term Control Plan Update, Supplemental Documentation*, Volume 3: *Basis of Cost Opinions*, September 2009. As of March 20, 2017: http://www.phillywatersheds.org/ltcpu/Vol03_Cost.pdf; "Green City, Clean Waters: Philadelphia's Long-Term Control Plan to Address Combined Sewer Overflows," October 24, 2013. As of March 20, 2017: http://www .montcopa.org/DocumentCenter/View/6567; Pittsburgh Water and Sewer Authority, "Greening the Pittsburgh Wet Weather Plan," July 2013. As of March 20, 2017: http:// apps.pittsburghpa.gov/pwsa/PWSA-Greening_the_Pittsburgh_Wet_Weather_Plan .pdf; Rossman, Lewis, *Stormwater Management Model User's Manual*, Version 5.1, Cincinnati, Ohio: National Risk Management, Laboratory Office of Research and Development, U.S. Environmental Protection Agency, EPA/600-R-14/413b, revised

quality by removing nitrogen from fertilizers or pollutants from roads and parking lots.[142] GSI mimics natural processes through a number of human inventions, such as cuts in a street curb to direct runoff into vegetated areas and bioretention or infiltration trenches.[143] In Syracuse, in 2011, Concord Place became the city's first "green street,"[144] created in part to manage stormwater by the installation of infiltration trenches along the street corridor. Stormwater enters the system through the existing storm drain connections in the street. Instead of the collected water flowing to the sewer system, as was previously the case, the water is directed to an underground trench filled with a stone base. As the water enters the trench, it slowly filters through the compacted stone and soil, eventually releasing into the groundwater.[145]

Stormwater planters[146] are another kind of structure to manage stormwater. Excess runoff is directed into an overflow pipe connected to the existing combined sewer pipe. The City of Philadelphia has installed a number of stormwater planters. Of particular note is the one at Columbus Square, being the first of its kind to be installed by the Philadelphia Water Department, converting a portion of Reed Street into a "green street."[147] The city offers grants to landowners to employ stormwater management systems infrastructure.[148]

September 2015. As of March 20, 2017: http://nepis.epa.gov/Exe/ZyPDF.cgi?Dockey =P100N3J6.TXT.

[142]   Fischbach *et al.*, *supra* note 77, at 18.

[143]   Cities of Mesa and Glendale, AZ, have incorporated these techniques in their Low-Impact Development toolkit. http://www.mesaaz.gov/home/showdocument?id= 14999.

[144]   https://savetherain.us/wp-content/uploads/2011/08/Concord-Place.pdf.

[145]   In addition to the underground infiltration system, Concord Place also received a new mill and pave application to the street surface, paid for by the City of Syracuse. This type of project is unique among green infrastructure projects–although above the surface, it appeared to be a traditional street paving process, below the street, green infrastructure was installed to manage stormwater more effectively and protect water resources. The completion of the renovation of Concord Place is the first of several planned "green street" projects within the "Save the Rain" program.

[146]   They are lined with permeable fabric, filled with gravel or stone, and topped off with soil, plants, and sometimes trees. The top of the soil in the planter is lower in elevation than the sidewalk, allowing for runoff to flow into the planter through an inlet at the street level.

[147]   http://archive.phillywatersheds.org/what_were_doing/green_infrastructure/ tools/stormwater-planter.

[148]   StormWaterIncentives progam, https://www.phila.gov/water/wu/stormwater/ Pages/Grants.aspx.

E   Land Use Regulation

The essential measures for adapting to a new climate must go beyond merely employing green technology, adjusting thermostats, and controlling pollution, but will require a rethinking about the underlying form of our communities. While it is doubtful that climate considerations drove the early planners and designers as they envisioned the city's form, in the wake of climate change, the climate must form a basic determinant of design. Climate must inform the "structural, environmental, economic, social, organizational, visual criteria of design."[149] Land use must be climate-cognizant. Urban micro-climatology should manifest itself in urban design decisions since climate is very much connected to a location in terms of temperature, moisture content of the air, rain, wind, fog, snow, insolation, cloudiness, and general air quality.[150] This means that certain areas should be shaded at particular times, buildings should be required to achieve a particular level of energy performance, streets should be oriented to facilitate traffic flow and airflow in buildings, public buildings should be located at accessible places, and pedestrian walkways should be tree-canopied.[151]

Climate-cognizant design is also informing the rate of allowable development. More land for increasing housing demands is being found through infill and redevelopment, not sprawl.[152] Growth controls are being employed to help cities achieve resiliency and sustainability. They contain existing density, thereby reducing disaster risk, particularly in low-lying coastal areas, along with major earthquake faults and major rivers.[153] Developers are being required

---

[149]   *Id.*

[150]   Evyatar Erell, David Pearlmutter, and Terrence Williamson, Urban Micro Climate: Designing the Space Between Buildings 5, 15–17 (2012).

[151]   Congress for the New Urbanism, https://www.cnu.org/resources/what-new-urbanism.

[152]   *See* Robert L. Liberty, *Ninth Annual Norman Williams Distinguished Lecture in Land Use Planning and the Law, February 7, 2013: Rising to the Land Use Challenge: How Planners and Regulators Can Help Sustain Our Civilization,* 38 Vt. L. Rev. 251, 257 (2013) ("The essence of most residential zoning, from the time of its inception a century ago, is the use of the state's police powers to separate housing by its type and cost and thereby segregate the residents by their income, and by extension, their race, ethnicity, and national origin"). Liberty, *supra* note 55, at 260.

[153]   Lisa Grow Sun, *Smart Growth in Dumb Places: Sustainability, Disaster, and the Future of the American City,* 2011 BYU L. Rev. 2157, at 2166-2167 (2011). Sun recounts that some commentators have previously identified urbanization as a factor in disaster risk. *Id.* However, the relationship between urbanization and disaster risk is likely more complicated than has sometimes been assumed. Population density can be seen both to exacerbate (through the high percentage of impervious surfaces, urban heat island effect, and increase in evacuation time) and to mitigate (by multistory buildings

to build houses closer to each other and to sidewalks, integrate transit-oriented features, and conduce to sustainable communities.[154] The cutting down of trees on private land is being limited;[155] parks and opens space are being required.[156]

F   Water

According to the Intergovernmental Panel on Climate Change (IPCC),[157] much of the impact of climate change will be felt in the water sector. This impact will manifest itself on two fronts: water as a resource and water as a hazard. As a resource, the availability of good-quality water is the basis for the well-being of the ever-increasing number of people living inside cities. The confluence of growing urban populations and climate change-driven drought calls for careful management of water and of land uses that impact water resources. The demonstrable connections between land and water are receiving greater attention as local governments begin to understand how decisions about land use can dramatically impact water demand.

---

serving as refuges during flooding events) climate change effects. *Id.,* citing Michael MacRae, Tsunami Forces Debate Over Vertical Evacuation, Am. Soc'y of Mech. Eng'rs (Apr. 2011), http://www.asme.org/kb/news---articles/articles/manufacturing---processing/tsunami-forces-debate-over-vertical-evacuation (discussing the possibility of "vertical evacuation" to the higher floors of multistory buildings during tsunamis).

[154] John R. Nolon, Part 18, Zoning's Centennial, Zoning: Shaping and Attracting Economic Development, online site. Some of the techniques being employed towards these ends are fast-tracking the planning and rezoning of downtowns, offering density bonuses, and creating traffic improvements; infill development, and creative development of open spaces adjacent to corporate, medical, educational, and non-profit buildings; adopting the USGBC's LEED-ND standards; and zoning to allow scattered sites throughout the neighborhoods within walking distance of train stations. *Id. See also* John R. Nolon, *An Environmental Understanding of the Local Land Use System,* 45 ELR 10215, 10219–10220, 10224, 10234 (2015) (discussing clustering, planned unit development and preservation districts). Homes must be removed from flood-prone areas. Local Gov'ts For Sustainability USA, Local Governments, Extreme Weather And Climate Change 2012 5 (2012), *available at* http://www.resilientamerica.org/wp-content/uploads/2013/06/ICLEI_extreme_weather_cities_fact_sheet_2012.pdf (describing plan adopted in King County, Washington to demolish chronically flooded homes).

[155] Trees must be allowed to stand. Pub. Res. Code § 25980, 25984 (trees planted before the installation of solar collectors are protected).

[156] *See e.g., New Jersey Shore Builders v. Township of Jackson,* 972 A.2d 1151 (N.J. 2009) (upholding a local ordinance that prescribed taking down trees on private property or requiring an in-lieu payment into a fund); *see generally* Nolon, *supra* note 154, at 10220, 10223.

[157] IPCC Fifth Assessment Report, https://www.ipcc.ch/assessment-report/ar5/.

Strategies that impact urban form, such as compact development, infill development, or smaller lot sizes, can drive down water use when compared to single-family homes on large lots. Pervious cover and green infrastructure can help direct stormwater and other runoff into water recharge areas. The California Adaptation Plan calls for a 20% reduction in per capita water use.[158] Texas, having recently suffered from drought, has called for a coordinated response through the National Integrated Drought Information System (NIDIS).[159] The Integrated Regional Water Management Plan for the Greater Los Angeles County (GLAC)[160] includes water desalination, groundwater management, combined use of surface water and groundwater, water storage, improved water conservation and efficiency of urban water use, water recycling, and water transfers from different regions in the state.[161]

The Colorado Water Plan includes the use of technologies for monitoring water availability and for profiling atmospheric changes for predicting water levels. The plan also contains conservation and management strategies for all water providers and provides funding, technical support, and training workshops to assist water providers in improving the management of their water systems, including techniques such as water budgets, smart-metering, comprehensive water loss management programs, savings tracking and estimating tools, and improved data collection on customer water uses. The plan contemplates adopting higher-efficiency technologies and provides incentives to water providers to retrofit inefficient systems with efficient ones and encourages more efficient irrigation systems. The objective is to reduce projected 2050 demands by 400,000 acre-feet through active conservation savings. This goal will be achieved through landscape and irrigation ordinances, green infrastructure ordinances, and more stringent green-construction codes that include higher efficiency fixtures and appliances and water-wise landscapes. The plan urges the removal of barriers to green-building and infrastructure and work with the appropriate agencies to adapt regulations to allow for graywater, green infrastructure, on-site water recycling, and other

---

[158] National Climate Assessment Report, [hereinafter NCA-3], note 40, https://nca2018.globalchange.gov/.

[159] *Id. at* note 48.

[160] http://www.ladpw.org/wmd/irwmp/.

[161] The State of California faced an extreme drought that required a declaration of a State of Emergency by Governor Brown on January 17, 2014. Enforceable water conservation measures were one of the key strategies. Governor Brown sought to reduce water usage by all Californians by 20% to confront this emergency. Accessed August 17, 2015:

http://www.waterboards.ca.gov/board_decisions/adopted_orders/resolutions/2014/rs2014_0038_regs.pdf; http://www.sacbee.com/2014/07/29/6591112/new-statewidewater-waste-prohibitions.html.

aspects of green developments. Overall, the plan makes plain that any water use should become a part of the larger land use plans, including expediting permitting for high-density buildings and developments that incorporate certain water efficiency measures, such as efficient irrigation systems; including water supply and demand management in comprehensive plans; installing climate-appropriate landscapes; and understanding the societal and environmental benefits of urban landscapes.[162]

To limit water waste, cities are employing metered water use;[163] advanced plumbing technologies;[164] and filtration by soil and roots runoff capture systems.[165]

G   Disaster Management and Resiliency

Climate impacts are on the rise and must be planned for and reckoned with. It is imperative to be ready to respond to protect and save lives and community when the next severe event occurs. And, cities must be resilient, i.e., possess the capacity to rebound, positively, adapt to, or thrive amidst changing conditions, or challenges, including disaster and climate change, in order to maintain quality of life, healthy growth, durable systems, and conservation of resources for present and future generations.[166] In neither of the two most severe recent weather events—Hurricane Katrina and Superstorm Sandy—were the cities hardest hit prepared for the onslaught or aftermath. But, from both, there were stern lessons learned of the need to plan for disasters. We learned that it is critical to set up paths and mechanisms for communicating with residents, evacuating them, and for delivering food and medical care. The League of American Cities has developed guides for preparedness for a range of disasters, from biohazard to wildfires.[167] Many local governments have taken the lessons from Sandy and Katrina to heart and have developed comprehensive disaster preparedness plans.[168] The components of an effective disaster management plan include the creation of centralized police, fire, and emergency medical service units that cater to unique vulnerabilities or designated zones; a triage

---

[162] *Id.*

[163] PlaNYC, at 27.

[164] *Id.*

[165] *See e.g.,* City of Portland, https://www.portlandoregon.gov/bes/article/188636.

[166] *See e.g.,* The Colorado Resiliency Framework, https://sites.google.com/a/state .co.us/coloradounited/resiliency-framework.

[167] https://www.nlc.org/topics/public-safety/disaster-preparedness.

[168] New York Rising Community Reconstruction Plans, https://stormrecovery.ny .gov/nyrcr/final-plans, *see generally* Rockefeller Foundation, https://www.rockefeller foundation.org/blog/six-great-ideas-from-the-national-disaster-resilience-building -competition.

system for customers during water and power shortages; building local area communications hubs with protocols for communications; elevating communications equipment to roofs of buildings; and encouraging backup generators and power sources for businesses to stay open. The New York City Rising Plan allocates funding for health and social service providers to make building-level capital upgrades to ensure continuity of service during and after an emergency through the critical facility upgrades program.[169]

H   Transportation

Transportation systems are being upgraded, and cities are investing in measures to facilitate less polluting means of movement, installing charging stations for electric cars,[170] and facilitating biking and walking as desirable modes of transportation.[171]

I   Economic Incentives for Adaptation Measures

No doubt, converting vacant areas into greenspace, community gardens, rain gardens, and other assets produces many public benefits. But many of these parcels were under threat for redevelopment due to their prime location or topography. While conversion reveals the natural beauty of trees and flowers, improves air and water quality, facilitates healthy food production, and promotes a sense of community, once converted to public use, the land is no longer taxable. At the same time, if these parcels remain taxable, a nonprofit community organization may be reluctant to acquire them because of the additional financial burden of paying property taxes. One response to this is the use of transferable development rights (TDR).[172] By severing development rights from a parcel dedicated to public use, those development rights stay in the marketplace and allow the parcels to continue serving as a source of tax revenues in a way otherwise not realizable through standard zoning methods. There are two primary forms of TDR programs in current use, but in either form, the first step is for the local governing body to establish the program in accordance with any existing planning code. In the most common program, the landowner can sell the development rights from their property directly to a developer through

---

[169] *Id.*

[170] 2014 Cal. Stat. s. ch. 529.

[171] *See e.g.,* City of Denver, https://denver.streetsblog.org/2016/04/28/denver-city -council-looks-to-create-fund-for-a-complete-sidewalk-network/.

[172] Using Transfer Development Rights to Facilitate and Sustain Community Green Space and Gardens Roy Kraynyk – Allegheny Land Trust, https://alleghenylandtrust .org/wp-content/uploads/2017/09/20170919_TDRWhitepaperv2.0.pdf.

a negotiated sales contract. The developer then uses those development rights to increase the density of houses or commercial square footage on another piece of property under the program regulation. A second method involves a development rights bank set up by the local government. In this method, the local government would purchase development rights from properties in areas that it wants to protect from development or green infrastructure development. Developers, who wish to develop at a higher density or more square footage than current zoning allows, would then purchase development rights from the local government. The local government then uses the revenue to purchase additional development rights from other properties it desires to protect. In Pennsylvania, nearly three dozen municipalities have incorporated the TDR tool into their zoning ordinances.[173]

Local governments have employed a host of other economic instruments and packages to achieve buy-in or force adherence with climate adaptation goals. These instruments include both taxes and their theoretical opposite, subsidies. Taxes and charges are being assessed based on the level of GHG emissions, the idea being that the incentive to avoid taxes will channel energy usage to lower levels. Some cities assess fees based on water usage.[174] North Carolina assesses a real estate recording fee.[175] New York imposes a surcharge on electric bills for a green bank.[176] The District of Columbia imposes a fee on plastic and paper bags to defray the costs of installing green roofs.[177]

On the subsidies side, there are tax credits for homeowners who install efficient outdoor landscapes and irrigation as part of the integrated funding plan. Water efficiency grants are available to homeowners.[178] Stormwater management retrofit grants are available.[179]

---

[173] Sections 603 (c) (2.2) and 619.1 of the PA Municipalities Planning Code.

[174] *See e.g.*, Boston, http://www.bwsc.org/sites/default/files/2019-02/2019%20Rate%20Doc.pdf; New Mexico, https://www.abqjournal.com/577825/rates-are-high-but-usage-is-low.html. Santa Fe water rates are among the highest, but usage is low. In Santa Fe, a water bill would be $144 for the same amount of water that Las Vegas residents pay $41.50 for on their average monthly bill.

[175] *See* https://files.nc.gov/ncdeq/climate-change/resilience-plan/2020-Climate-Risk-Assessment-and-Resilience-Plan.pdf pp. 4-13.

[176] *See* New York Green Bank, https://greenbank.ny.gov/

[177] Anacostia River Clean Up and Protection Act of 2009 (places a five-cent fee on disposable plastic and paper bags provided by any District retailer selling food or alcohol.); *see also* https://www.nrdc.org/sites/default/files/RooftopstoRivers_WDC.pdf

[178] *See e.g.*, N.Y. https://www1.nyc.gov/site/dep/news/19-067/funding-available-new-york-city-property-owners-who-conserve-water; California: https://californiaseec.org/energy-efficiency-funding-opportunities/.

[179] https://www.phila.gov/water/wu/stormwater/Pages/Grants.aspx; Greened Acre Retrofit Program Grant – https://www.phila.gov/water/wu/Stormwater%20Grant%20Resources/GARPFactSheet.pdf.

All adaptation programs work best when citizens buy into them. This is why adaptation plans and initiatives, almost invariably, contain a stakeholder component. They explain the imperatives of action but also appeal to a larger sense of community.[180]

---

[180] One notable strategy is from Boulder County, Colorado, after the Big Thompson Canyon Flood of 2013. The Long-Term Flood Recovery Group (LTFRG) partnered with the City of Boulder, Boulder County, Lyons, Jamestown, Longmont, nonprofits, the education and business community, and faith-based groups to launch BoCo Strong – a series of community conversations to help Boulder County strengthen its resilience following that flood. http://bocofloodrecovery.org/conversations-on-building -resilience-into-our-communities-scheduled-throughout-boulder-county. The Boulder County plan called for a series of "informal, neighborhood-style meetings will take place throughout July and early August at multiple venues in an effort to capture important lessons learned from the 2013 Flood" and sought to learn "what worked for you during the flood? What didn't work for you? What preparations have you or your community made to prepare for other disasters such as wildfire or drought? What makes households, neighborhoods and communities resilient?" The plan explained that "[c]ommunity input [would] be used to develop Resilience Best Practices that would be available to all residents of the county. In addition, the findings would be presented to community leaders from all sectors (government, business, nonprofit, faith, education, etc.) identifying what we should work on to enhance our resilience." *Id.*

5. Litigating government (in)action on climate change

Katrina Fischer Kuh

Litigation brought under a variety of sources of law, including the common law, statutes, and constitutions, to prompt governments to undertake mitigation and/or adaptation has proved to be an important means to define the responsibilities and shape the responses of governments relating to climate change. This litigation is particularly important in the United States (US) because of the absence of sustained political will and federal law aimed specifically at reducing US greenhouse gas (GHG) emissions. Indeed, without catalyst litigation, the US would have little federal "law" being brought to bear on climate change at all. However, even in jurisdictions like the Netherlands with relatively robust emission reduction commitments and laws, climate advocates have successfully petitioned the courts to heighten government responsibilities relating to climate change. Moreover, litigation to force government action on climate change has, itself, become an important component of the law of climate change.

This chapter describes important catalyst litigation that has helped to define the content and scope of government obligations related to climate change. It begins by reviewing the use of litigation to compel climate mitigation and/or adaptation under laws of general application, including most notably the Clean Air Act (CAA) and the National Environmental Policy Act (NEPA). It then highlights significant efforts to impose extra-statutory mitigation or adaptation obligations on governments, i.e., obligations grounded in the common law, constitutions, and/or human rights law and enforced through domestic courts. This review of catalyst litigation aimed at compelling government action on mitigation and/or adaptation reveals that litigation has been an important tool to activate and enforce climate change-relevant statutes and suggests both the promise and difficulties of using litigation grounded in extra-statutory sources of law to prompt government action on climate change.

Focusing narrowly on current mitigation or adaptation requirements imparts a snapshot of climate change law that will quickly be eclipsed by new developments as the law of climate change evolves. To make sense of the existing patchwork of laws governing climate mitigation and/or adaptation, particularly

109

in the US, it can be helpful to understand the origins of the laws and how their implementation has been shaped through litigation. Knowing not just how a statute is presently interpreted or how the scope of government obligations is presently defined, but also the litigation and arguments giving rise to those understandings (or raised to challenge them), can help to predict future directions in the development of climate change law and provide perspective on the significance and meaning of those developments when they occur. The account provided in this chapter thus focuses on how litigation has shaped climate change law to date and has the potential to shape it going forward.

I   COMPELLING GOVERNMENT CLIMATE ACTION UNDER LAWS OF GENERAL APPLICATION

The use of laws of general application—that were not specifically designed to address climate change—to respond to climate change has proved particularly important in the US because of the absence of political will to adopt federal climate change legislation or otherwise take strong federal action to reduce emissions. These laws hold out the possibility of federal action on climate change even in the absence of political agreement between Congress and the president. Federal agencies acting at the direction of the president typically decide in the first instance whether and how laws of general application are applied to climate change. Those agency interpretations are then subject to court review to assess whether they reside within the permissible bounds of the statute. Under a presidential administration hostile to federal climate efforts, an agency will be inclined to interpret the statute to require minimal action or controls; under an administration supportive of federal climate efforts, an agency will be inclined to interpret the statute to support robust federal action on climate change. Courts are then put in the position of looking to the statute to determine whether it requires more action than the agency has proposed (hostile administration, statutory minima) or in fact gives the agency the authority to take the strong climate action it is proposing (supportive administration, statutory maxima). Litigation in this context thus often explores permissible statutory minima and maxima.

The most important law of general application for addressing climate change in the US is the CAA. The CAA is the primary federal law designed to control the emission of pollutants to air to protect the public health and welfare and, as described in Chapter 2, agencies and courts interpret the general language of select provisions of the CAA to support, and in some cases require, the regulation of GHG emissions. Notably, the relevant provisions of the CAA are generally applicable, i.e., they are not specifically directed to climate change or GHG emissions, but oriented toward reducing air pollution more generally. With broad reach, well-developed experience with and infrastructure for implementation, and substantial success at reducing ambient levels of numerous air pollutants, there would seem to be much to recommend the CAA as the anchor for GHG emission controls in the US. There are, however, recognized difficulties with relying on the CAA to control GHG emissions. Perhaps most importantly, the CAA's structure anticipates that state reductions in emissions of a pollutant can achieve measurable local or regional progress toward a national ambient air quality standard; GHG emissions worldwide mix in the atmosphere, rendering a single state largely unable to discernibly impact ambient concentrations of GHGs. Additionally, there is some uncertainty about whether the CAA permits the use of flexible mechanisms (cap and trade, carbon taxes, etc.) that might maximize efficiency by decreasing compliance costs. And, despite its broad reach, the CAA does not cover some important sources of GHGs (for example, fossil fuel extraction like drilling for oil or mining for coal). Many sophisticated actors—from climate activists to regulated entities—express a preference for regulation of GHG emissions under climate change-specific legislation as opposed to under the CAA. However, in the absence of political support for new federal climate change legislation, efforts to control GHG emissions at the federal level have centered on the CAA occasioning numerous legal disputes about whether and how the statute can be understood to require or support climate change regulation. Chapter 2 provides an overview of the regulation of GHGs under the CAA. The analysis in this chapter supplements that overview by focusing on the role of catalyst litigation in the evolution of GHG regulation under the CAA.

In *Massachusetts v. EPA*, the seminal case that opened the door for the regulation of GHG emissions under the CAA, a group of states, local governments, and private organizations petitioned the EPA to initiate rulemaking under section 202(a)(1) of the CAA to limit vehicle tailpipe emissions of GHGs.[1] Section 202(a)(1) requires the EPA to set standards for motor vehicle emissions of "any air pollutant .... Which in [the Administrator's judgment] cause or contribute to air pollution which may reasonably be anticipated to endanger the public health or welfare."[2] The petition for rulemaking contended that this statutory charge required the EPA to set tailpipe emission standards for GHGs. The EPA denied the petition, consistent with the policy preference of then-President George W. Bush. The EPA explained that it did not possess authority under the CAA to regulate GHGs because the term "air pollutant" in the CAA could not be understood to include GHGs. The EPA reasoned that Congress did not intend to regulate GHGs under the CAA because the structure

---

[1]   Massachusetts v. E.P.A., 549 U.S. 497 (2007).
[2]   42 U.S.C.A. § 7521(a)(1) (West 2020).

of the CAA was not readily amenable to the regulation of GHGs, actions taken by Congress after the enactment of the CAA signaled that Congress did not believe that GHGs were already governed by the CAA, and that there should be a clear statement of congressional intent to invoke the CAA's broad provisions to regulate GHGs because to do so would have a significant impact on the CAA and economy. The EPA explained further that, even if it did possess the statutory authority under the CAA to regulate GHGs as air pollutants, it would not do so for a variety of reasons, including that science about anthropogenic contributions to climate change remained too uncertain, regulating GHGs under the CAA would conflict with ongoing independent efforts to control GHG emissions (including the negotiation of an international climate change agreement), and the CAA was not well suited to efficiently regulate GHGs. The petitioners then sought court review seeking a court order that the CAA compelled the EPA to regulate GHG emissions under section 202(a)(1).

As described in Chapter 2, the Supreme Court ultimately held that GHGs are potentially subject to regulation as air pollutants under the CAA and directed the EPA to provide a rationale for declining to regulate GHGs under section 202(a)(1) that was grounded in the statute. The decision had significant and specific practical implications under the CAA that flowed from its forceful nudge to the EPA to initiate a series of regulatory actions under the CAA, eventually leading to the regulation of some GHG emissions under the CAA (including tailpipe emissions, emissions from new or modified electric generating units, and emissions from new or modified major sources of non-GHG pollutants). The decision also laid the groundwork for arguments that other provisions of the CAA require reductions on emissions from still more sources, including existing electric generating units. Significant, existing regulation of GHGs under the CAA as well as prospects for future regulation are summarized in Chapter 2.

Massachusetts v. EPA's influence on the development of nascent climate change law and litigation in the US went far beyond its relatively limited holding, immediate precedential command, and subsequent implications for regulation under the CAA. The case established basic precepts that have supported much subsequent climate change litigation and regulation in the US. These precepts are now well established but were very much open questions at the time of *Massachusetts v. EPA* and, had they been rejected or approached in a different way by the Supreme Court, could have significantly impeded the use of litigation to compel government action on climate change. Specifically, the Supreme Court (1) established that the diffuse and generalized harms, numerous contributors, and long timeframes associated with climate change would not necessarily defeat litigants' standing to press claims grounded in climate change claims in court; (2) declined to allow the EPA to use the complexity and asserted uncertainty of climate change science as a shield for inaction; and (3) recognized the potential for laws of general applicability, adopted without the specific intent to address climate change, to be interpreted to encompass climate change within their mandates.

A litigant can only press a climate change-related claim in court if it has standing, which requires showing that it has suffered a concrete and particularized injury that is either actual or imminent, that the injury is fairly traceable to the defendant, and that it is likely that a favorable decision will redress that injury.[3] Injury related to climate change presents significant difficulties with respect to making these required showings as it may prove difficult to link specific injury to climate change, injury may be projected to occur over long and uncertain timeframes, and the large number of contributors to climate change and the global nature of the problem may make it difficult to assign blame to particular defendants or show that a court remedy directed to particular defendants will redress the injury. Only one litigant needs to establish standing for a case to proceed, and in *Massachusetts v. EPA*, the Supreme Court focused its analysis on whether the State of Massachusetts possessed standing. The Supreme Court reasoned that the plaintiffs' affidavits indicated that sea-level rise was already occurring and causing injury to Massachusetts' coastal lands and that even "a small incremental step" toward reducing GHG emissions could "slow or reduce" global warming.[4]

In holding that Massachusetts possessed standing, the Supreme Court laid to rest the objection that the nature and scope of climate change—the long timeframes, broadly shared harms, and multiple contributors—necessarily render climate change-related disputes unfit for judicial resolution. Concluding (over a vigorous dissent) that Massachusetts had standing, the Supreme Court reasoned that none of these issues defeated standing. To be sure, uncertainty remains after *Massachusetts v. EPA* about the precise contours of standing to litigate climate change-related harms. The Court found that Massachusetts' status as a sovereign state and the procedural nature of the claim relaxed the demands of redressability and immediacy,[5] asserting that "[g]iven that procedural right and Massachusetts' stake in protecting its quasi-sovereign interests, the Commonwealth is entitled to special solicitude in our standing analysis."[6]

---

[3]   Lujan v. Defenders of Wildlife, 504 U.S. 555, 560–61 (1992).

[4]   Massachusetts v. E.P.A., 549 U.S. at 524–25.

[5]   *Id.* at 518 ("When a litigant is vested with a procedural right, that litigant has standing if there is some possibility that the requested relief will prompt the injury-causing party to reconsider the decision that allegedly harmed the litigant.") ("We stress ... the special position and interest of Massachusetts. It is of considerable relevance that the party seeking review here is a sovereign State and not, as it was in Lujan, a private individual.").

[6]   *Id.* at 520, 1454–55.

It remains contested how the absence of those factors supporting special solicitude (as when a private litigant brings a claim under the common law) affects
the standing analysis in climate change cases. As discussed further in Chapter
7, the Supreme Court subsequently split on whether a state possessed standing
to sue large emitters under common law nuisance in *American Electric Power
Co. v. Connecticut,* and some lower courts have dismissed climate change
cases for lack of standing.[7] By and large, however, after *Massachusetts v. EPA*
courts have regularly found that plaintiffs in climate change-related suits can
establish standing despite the temporal, geographic, and numerous-contributor
attributes of climate change.

In addition to affirming the ability of at least some plaintiffs to bring claims
grounded in climate change, *Massachusetts v. EPA* also signaled that courts
would not invoke the complexity of climate change science to reflexively
dismiss plaintiffs' claims or defer to agencies. Overall, judicial fora have
proved to be inhospitable to weakly supported, politicized mischaracterizations
of the science of climate change, and *Massachusetts v. EPA* stands as an early
and influential example of judicial competence and independence with respect
to evaluating climate change science. Recall that one of the chief rationales
offered by the EPA for declining to grant the petition for rulemaking was its
assertion that there was residual uncertainty about the connection between
anthropogenic GHG emissions and climate change. While the Supreme Court
did not directly reject the EPA's assertion, it reviewed in detail the development of scientific learning about climate change and, instead of deferring
to the EPA's policy-inflected characterization of climate change science,
observed that the EPA had selectively cited to particular conclusions within
a report prepared by the National Research Council and reasoned that the EPA
could not "avoid its statutory obligation by noting the uncertainty surrounding
various features of climate change," instructing that "[i]f the scientific uncertainty is so profound that it precludes EPA from making a reasoned judgment
as to whether greenhouse gases contribute to global warming, EPA must say
so."[8] The Supreme Court declined to defer to the EPA's characterization of
climate change science; one persuasive explanation offered by scholars for the
Supreme Court's non-deferential posture is that the Supreme Court was sensitive to high-profile incidents suggesting that the EPA's approach to climate
change science under President George W. Bush was politicized—skewed

---

[7]   Washington Envtl. Council v. Bellon, 732 F.3d 1131, 1142-43 (9th Cir. 2013)
(distinguishing *Massachusetts v. EPA* and dismissing environmental organizations'
claim seeking to require Washington to impose GHG limits on five oil refineries under
the CAA because the connection between the emissions and plaintiffs' harm was too
attenuated).
[8]   Massachusetts v. E.P.A., 549 U.S. at 534.

by the administration's opposition to federal regulation of GHG emissions.[9] Through this independent approach to climate change science, the Court signaled that the judiciary would not allow deference doctrines and scientific complexity to shield agency decision-making from court oversight and, more broadly, that the courts would be inhospitable fora for manufactured debates about core principles of climate change science, including whether climate change is occurring and whether it is anthropogenic.

Finally, the Court in *Massachusetts v. EPA* rejected the view that climate change is too exceptional to fall within the scope of statutes of general application. *Massachusetts v. EPA* came to the courts in the posture of an agency, delegated authority to implement the CAA and doing so consistent with the policy preferences of an administration hostile to federal action on climate change, interpreting the CAA in a miserly fashion to afford no authority to regulate GHG emissions. A key part of the EPA's narrative—underlying its effort to invoke to the major questions doctrine, infusing its explanation of the significance of post-enactment legislative history, motivating its comparisons of climate change to ozone depletion—was the idea that climate change is so exceptional (scientifically complex, global and significant in terms of both direct impacts and the impacts of mitigation measures on the economy and society) that Congress could not have possibly intended for it to fall within the scope of statutes of general application. In other words, general language in a statute otherwise applicable to climate change by virtue of its plain meaning must be understood by courts to except climate change. The Supreme Court soundly rejected this argument, reasoning:

> While the Congresses that drafted § 202(a)(1) might not have appreciated the possibility that burning fossil fuels could lead to global warming, they did understand that without regulatory flexibility, changing circumstances and scientific developments would soon render the Clean Air Act obsolete. The broad language of § 202(a)(1) reflects an intentional effort to confer the flexibility necessary to forestall such obsolescence.[10]

The Supreme Court thus treated climate change as encompassed within the scope of the environmental challenge that the CAA's broad language was intended to address, even though the CAA was not specifically adopted or designed to address climate change and the relevant provision did not reference climate change specifically. This approach normalized climate change within the context of statutory interpretation in a fundamental way, discour-

---

[9]   Jody Freeman & Adrian Vermeule, *Massachusetts v. EPA: From Politics to Expertise*, 2007 SUP. CT. REV. 51, 51–67 & 92–96 (2007).

[10]   Massachusetts v. E.P.A., 549 U.S. at 532.

aging if not entirely foreclosing arguments that it was necessary for climate change to be specifically contemplated at the time a statute was adopted to be understood to reside within its auspices.

Massachusetts v. EPA opened the door to the use of statutes of general application—including NEPA, the Endangered Species Act, and the Securities Exchange Act—to address climate change. Legal disputes about how such statutes must or can be applied in the context of climate change constitute a large and significant portion of domestic climate litigation. As noted above, the framing of these legal disputes has tended to depend upon whether the presidential administration guiding the regulatory progress is hostile to or supportive of federal regulation of climate change. Litigants ask courts to evaluate either whether a hostile administration's interpretation of the statute fails to satisfy minimum statutory requirements (*Massachusetts v. EPA*) or, alternatively, whether a supportive administration's interpretation of the statute exceeds its authority (*Utility Air Regulatory Group v. EPA*, discussed in Chapter 2). Apart from the CAA, some of the most significant disputes about the application of a law of general application to climate change have arisen under NEPA and similar environmental review statutes.

NEPA requires that federal agencies evaluate the environmental impacts of major federal actions significantly affecting the quality of the human environment. NEPA is a statute of general application (it does not specifically reference climate change) and it is also a procedural statute—it can require the consideration of climate change in environmental review but it does not (and cannot) directly compel agencies to take particular mitigation or adaption actions. However, environmental review with robust consideration of climate change has the potential to embed climate change into a wide range of decision-making, thereby powerfully shaping our response to climate change. As in the case of the CAA, for NEPA to live up to its potential to inform and promote mitigation and adaptation, it is first necessary to establish the legal bases for applying the statute to climate change.

And, again, as with respect to the CAA, whether and, if so, how NEPA has been understood to apply in the context of climate change reflects the push and pull of supportive and hostile presidential administrations. Neither the text of the NEPA statute nor the regulations promulgated by the Council on Environmental Quality (CEQ) to guide agency compliance with NEPA specifically reference climate change. However, NEPA requires that agencies prepare a "detailed statement" describing "the environmental impact of the proposed action" as well as "any adverse environmental effects which cannot be avoided."[11] Under President Obama, CEQ released final guidance in 2016

---

[11]   42 U.S.C.A. § 4332 (West 2020).

explaining that NEPA requires agencies to consider climate change in some contexts, including by quantifying a proposed agency action's projected direct and indirect GHG emissions and treating those emissions as a proxy for assessing potential climate change effects.[12] President Trump rescinded the Obama-era CEQ NEPA guidance on climate change and in June 2019 CEQ issued new draft guidance that, while often vague, signaled that agencies need do far less to comply with NEPA than indicated in the prior guidance;[13] President Biden, in turn, quickly rescinded the Trump-era draft guidance and is reviewing and updating the Obama-era guidance. The CEQ under President Trump also promulgated amendments to its NEPA regulations, which, if they survive legal challenge, could limit agency obligations to evaluate climate change; those amendments are the subject of legal challenge and under reconsideration by the Biden CEQ.

Courts have frequently been called upon to evaluate whether agency treatment of climate change in NEPA review satisfies the requirements of the statute. While some important questions about what NEPA requires vis-à-vis climate change remain uncertain, both because of a lack of uniformity in court decisions and the flux in the regulatory landscape, it is useful to recognize some important, baseline propositions that now seem well established. Agencies generally cannot decline to consider climate change in NEPA review on the grounds that climate change is not reasonably foreseeable or its effects too uncertain. Even where an action would avoid or reduce GHG emissions, further analysis of climate change impacts may be required, particularly where alternatives to the action could avoid or reduce emissions further. The inability to precisely quantify the extent to which GHG emissions associated with an action would contribute to climate change does not excuse consideration of climate change impacts. And, while it has yet to be determined what volume of GHG emissions produced by an action is sufficient to require consideration of the contribution of those emissions to climate change, cases demonstrate that the threshold is not so high as to render the consideration of climate change impacts rare or unusual in NEPA analysis.

These baseline propositions, anchored in the statutory text, suggest that, absent an amendment to the text of NEPA itself, consideration of climate change impacts in NEPA review will continue. Important questions remain,

---

[12] Memorandum from the Council of Environmental Quality on Final Guidance for Federal Departments and Agencies on Consideration of Greenhouse Gas Emissions and the Effects of Climate Change in National Environmental Policy Act Reviews (Aug. 1, 2016), https://ceq.doe.gov/docs/ceq-regulations-and-guidance/nepa_final_ghg _guidance.pdf [https://perma.cc/Y9Z7-FE46].

[13] Draft National Environmental Policy Act Guidance on Consideration of Greenhouse Gas Emission, 84 Fed. Reg. 30,097 (June 26, 2019).

however, about when and how climate change will be incorporated into NEPA reviews and what courts will determine constitutes the minimum that NEPA requires in this regard. Two areas where NEPA review has great potential to meaningfully influence mitigation and adaptation are with respect to actions involving fossil fuel supply policies and projects (mitigation) and the built environment (adaptation). In both contexts, litigation brought challenging as deficient environmental reviews that fail to consider climate change or to do so adequately is driving the development of more robust review practices and illustrates the importance of another area of climate change litigation under a statute of general application.

Federal agencies control significant aspects of fossil fuel supply in the US, including most notably through approvals related to the extraction (granting leases authorizing drilling or mining on public lands) and transportation (construction of pipeline or export terminals) of fossil fuels. Litigation is testing important questions about the review required under NEPA in this context, including whether downstream (and in some contexts upstream) emissions must be considered; whether such emissions can be quantified and, if so, how; and whether, in addition to project-specific review, NEPA compels more comprehensive programmatic assessment of climate change impacts from multiple leasing decisions and pipeline approvals.[14] In one of the most significant cases applying NEPA to a fossil fuel infrastructure project, *Sierra Club v. FERC*, the D.C. Circuit Court of Appeals held that the Federal Energy Regulatory Commission (FERC) was required to either quantify (or explain why it could not quantify) downstream emissions from a natural gas pipeline project for which it issued a certificate of public convenience and necessity authorizing construction.[15] The pipeline project would have provided over 1 billion cubic feet of natural gas per day to Florida Power & Light and Duke Energy Florida to generate electricity. Although the FERC argued that it would not be a legal cause of the downstream emissions and that it would be impossible to predict the exact quantity of those emissions, the D.C. Circuit held that GHG emissions from combustion of the natural gas into electricity were reasonably foreseeable indirect effects from the pipeline approval and that the FERC had legal authority to consider those emissions in deciding whether to grant the certificate in light of its broad statutory authority to balance public benefits and

---

[14]   For a thorough analysis of questions about the application of NEPA to fossil fuel projects, *see* Michael Burger & Jessica Wentz, *Downstream and Upstream Greenhouse Gas Emissions: The Proper Scope of NEPA Review*, 41 Harv. Envtl. L. Rev. 109 (2017); Michael Burger & Jessica Wentz, *Evaluating the Effects of Fossil Fuel Supply Projects on Greenhouse Gas Emissions and Climate Change Under NEPA*, 44 Wm. & Mary Envtl. L. & Pol'y Rev. 423 (2020).

[15]   Sierra Club v. FERC, 867 F.3d 1357, 1357 (D.C. Cir. 2017).

adverse effects. It also noted that the FERC need not calculate emissions precisely, as NEPA permits it to make predictions based on reasonable assumptions. The court also rejected the FERC's assertion that "the EIS was absolved from estimating carbon emissions by the fact that some of the new pipelines' transport capacity will make it possible for utilities to retire dirtier, coal-fired plants," as NEPA requires the consideration of even beneficial effects.[16]

Numerous other cases are quickly developing parameters about when environmental review under NEPA must consider upstream and downstream emissions with respect to fossil fuel extraction and transportation projects and how those emissions are to be estimated. An important emerging issue relates to the timing and scope of review for federal actions relating to fossil fuel supply—whether, in addition to project-specific review, additional programmatic review needs to be conducted. Presently, there is a lamentable analytical gap because "the federal government has never conducted a programmatic analysis to evaluate the cumulative effects of its leasing decisions or transport approvals on fossil fuel use and GHG emissions," which results in "a patchwork of project-level NEPA documentation that provides only pieces of insight on how federal decisions about fossil fuel supply infrastructure affect fossil fuel use and GHG emissions."[17]

NEPA could also contribute significantly to effective climate change policy by discouraging investment in projects and approaches that are vulnerable to climate change impacts or, alternatively, encouraging projects and approaches that are resilient. This might mean, for example, taking sea-level rise into account in evaluating a coastal highway project or the siting of a new hazardous waste storage facility. While it seems obvious that we should consider climate change impacts when designing projects, federal agencies have been slow to fully incorporate such considerations into NEPA review, suggesting a potential role for litigation to drive more robust implementation.[18]

Two issues can complicate efforts to deploy NEPA to require the consideration of climate change impacts on a project and thereby center resiliency in agency decisions. First, there is arguably some uncertainty about whether NEPA requires the consideration of environmental effects *upon* a project in addition to environmental effects *caused by* a project. In *Ballona Wetlands*

---

[16]   *Id.* at 344.
[17]   Michael Burger & Jessica Wentz, Evaluating the Effects of Fossil Fuel Supply Projects on Greenhouse Gas Emissions and Climate Change Under Nepa, 44 Wm. & Mary Envtl. L. & Pol'y Rev. 423, 427 (2020).
[18]   For a comprehensive overview and empirical assessment of the application of NEPA in this context, *see* Jessica Wentz, *Assessing the Impacts of Climate Change on the Built Environment: A Framework for Environmental Review*, 45 ENVTL. L. REP. NEWS & ANALYSIS 11015 (2015).

Land Trust v. Los Angeles, a California court, interpreting California's state environmental review statute (the California Environmental Quality Act (CEQA)), held that an environmental impact review did not require consideration of the effects of sea-level rise on a development because "the purpose of an EIR is to identify the significant effects of a project on the environment, not the significant effects of the environment on the project."[19] Similar questions could be raised with respect to NEPA, although there are strong arguments that climate change impacts on a project should be considered during NEPA review, because factoring in climate change is necessary to accurately describe the baseline environment of the project, in many cases climate impacts upon a project may, in turn, lead to environmental effects from a project (as, for example, where a hazardous waste site floods),[20] and because a key purpose of NEPA is to facilitate the wise use of resources (thus creating an imperative to avoid waste).[21] Second, it may be difficult to identify climate change impacts at a regional or local level so as to discern climate change impacts on a particular project. However, the CEQ regulations governing NEPA implementation expressly instruct agencies on how to proceed in the face of incomplete or unavailable information[22] and advances in climate modeling continue to improve the capacity to project localized climate change impacts such that the lack of precision in local climate modeling should not excuse analysis under NEPA.

This survey of NEPA's application to climate change again illustrates the important role of statutes of general application, and litigation to press their implementation, in shaping government mitigation and adaptation in the US. As with respect to the CAA, presidential administrations supportive of federal climate change action have pushed for robust interpretations of NEPA to require meaningful consideration of climate change in environmental reviews, while administrations hostile to federal climate change action have sought to minimize the same. In the face of these competing interpretations of NEPA's requirements, courts are defining through case law the minimum (and maximum) in terms of what the statute requires vis-à-vis climate change. Two of the most important areas where the scope of NEPA's application is evolving

---

[19]   Ballona Wetlands Land Tr. v. City of Los Angeles, 201 Cal. App. 4th 455, 473 (2011).

[20]   Notably, California courts have since clarified that review does appropriately consider how climate change could impact a project in ways that would, in turn, impact the environment.

[21]   42 U.S.C.A. §4332(2)(C)(iv), (v) (2020) (instructing that an EIS should consider "the relationship between local short-term uses of man's environment and the maintenance and enhancement of long-term productivity, and … any irreversible and irretrievable commitments of resources which would be involved in the proposed action should it be implemented.").

[22]   40 C.F.R. §1502.22 (2020).

through both regulatory action and litigation are with respect to how agencies evaluating actions related to fossil fuel supply (including fossil fuel extraction and transportation) must calculate and consider the emissions associated with such actions and whether and how agencies need to evaluate how climate change will impact projects. Achieving robust consideration of climate change by federal agencies in environmental review in these contexts could significantly benefit US climate change mitigation and adaptation policy.[23]

II   INVOKING OTHER (NON-STATUTORY) SOURCES OF LAW TO COMPEL GOVERNMENT ACTION ON CLIMATE CHANGE

In 2015, a group of experts in environmental, human rights, and international law surveyed the law to discern "the current obligations that all States and enterprises have to defend and protect the Earth's climate" and issued the Oslo Principles on Global Climate Change.[24] The Oslo Principles, which are accompanied by a detailed legal Commentary and were drafted in part to "help judges decide whether particular governments are in compliance with their legal obligations to address climate change,"[25] posit that states have a legal obligation to limit emissions to an amount sufficient to avoid two degrees of warming as divided between countries on a per capita basis. The drafters broadly described the legal basis for this conclusion as follows:

> No single source of law alone requires States and enterprises to fulfil these Principles. Rather, a network of intersecting sources provides States and enterprises with obligations to respond urgently and effectively to climate change in a manner that respects, protects, and fulfils the basic dignity and human rights of the world's people and the safety and integrity of the biosphere. These sources are local, national, regional, and international and derive from diverse substantive canons, including, inter alia, international human rights law, environmental law and tort law.[26]

---

[23]   Environmental review under other authorities in different jurisdictions (including by local and state entities) could likewise beneficially influence mitigation and adaptation policy. In many other jurisdictions, environmental review authorities have been amended to specifically require the consideration of climate change, rendering the interpretive issues somewhat less difficult. Important questions do, however, arise with respect to the application of environmental review statutes that specifically require the consideration of climate change.

[24]   EXPERT GROUP ON GLOBAL CLIMATE OBLIGATIONS, OSLO PRINCIPLES ON GLOBAL CLIMATE CHANGE OBLIGATIONS (Mar. 1, 2015), https://globaljustice.yale.edu/oslo -principles-global-climate-change-obligations.

[25]   *Id.*

[26]   *Id.* at 3.

The drafters' reference to a network of intersecting sources is apt. While it seems clear that states have an obligation to reduce emissions, it can be very difficult and complicated to locate that obligation within existing sources of law, let alone to make it enforceable, creating a challenge for litigants seeking to compel government action on climate change. Despite the challenging nature of these claims, cases brought by litigants to require more aggressive government action to mitigate or adapt to climate change than what is currently required under existing domestic (statutory) law have proliferated. Litigants invoke a wide variety of sources of law in these actions but typically claim that the government is required to take more aggressive action on climate change under human rights law and/or the common law and/or the constitution. Notably, these cases seek to impose obligations on governments that go beyond what is required under domestic statutory law. Three significant cases—*Urgenda Foundation v. Netherlands*, *Greenpeace Nordic Association v. Ministry of Petroleum and Energy*, and *Juliana v. US*—provide an illustrative overview of this body of litigation.

A   Urgenda Foundation v. Netherlands

In 2015, the climate advocacy group Urgenda Foundation and 900 citizen co-plaintiffs brought suit against the Dutch government alleging that its failure to commit to more ambitious mitigation targets violated provisions of the Dutch Constitution, the European Convention on Human Rights (ECHR), and the government's duty of care under the Dutch Civil Code. In 2007, the Netherlands was planning to reach a target of reducing GHG emissions 30% from 1990 levels by 2020. In 2011, however, the Dutch government weakened its GHG emissions reduction targets, announcing a new target of reducing emissions by 20% from 2005 levels by 2020 (a reduction of 16% in the non-Emissions Trading System (ETS) sector and 21% in the ETS sector) in line with European Union standards. The plaintiffs sought declaratory relief and an injunction to compel the Dutch government to reduce GHG emissions by at least 25% when compared to the emissions levels of 1990.

The District Court ruled for the plaintiffs and ordered that the Dutch government aim to achieve reductions of emissions of at least 25% by the end of 2020 relative to 1990.[27] The initial portion of the District Court's decision did

---

[27]   Urgenda Found. v. Neth., HA ZA 13-1396 The Hague Dist. Ct. (Chamber for Comm. Affairs June 24, 2015) [hereinafter Urgenda District Court decision], *affirmed*, State of the Netherlands v. Urgenda Found., ECLI:NL:HR:2019:2007, Judgment (Sup. Ct. Neth. Dec. 20, 2019) (Neth.) [hereinafter Urgenda Supreme Court decision].

not bode well for the plaintiffs. The District Court began by reviewing many of the sources of law invoked by the plaintiffs and concluding that

> a legal obligation of the State towards Urgenda cannot be derived from Article 21 of the Dutch Constitution, the "no harm" principle, the UN Climate Change Convention, with associated protocols, and Article 191 of the Treaty on the Functioning of the European Union (TFEU) with the ETS Directive and Effort Sharing Decision based on TFEU.[28]

The District Court went on, however, to find that the government had a duty grounded in the Dutch Civil Code's doctrine of hazardous negligence to act "with due care towards society."[29] The District Court then looked to both hazardous negligence jurisprudence and to the authorities previously dismissed as not adequate to impose an independent legal duty to Urgenda to define the scope of the government's duty under Dutch negligence law.[30] Upon extensive review of those authorities, the District Court concluded that the Dutch government "acted negligently and therefore unlawfully towards Urgenda by starting from a reduction target for 2020 of less than 25% compared to the year 1990."[31]

The government appealed on 29 grounds to the Hague Court of Appeals and Urgenda brought its own cross-appeal, disputing the District Court's decision that it could not invoke Articles 2 and 8 of the ECHR. The Hague Court of Appeals and Dutch Supreme Court affirmed the order but located the duty of the Dutch government in the ECHR as opposed to the Dutch law of hazardous negligence.[32] Article 2 of the ECHR protects the right to life and has been interpreted by the European Court of Human Rights (ECtHR) to impose a positive obligation on contracting states "to take appropriate steps to safeguard the lives of those within its jurisdiction," including with respect to real and immediate risks posed by hazardous industrial activities and natural disasters.[33] Article 8 of the ECHR protects the right to respect for private and family life and "encompasses the positive obligation to take reasonable and appropriate measures to

---

[28]   Urgenda District Court decision ¶ 4.52.

[29]   Urgenda District Court decision ¶ 4.54.

[30]   Urgenda District Court decision ¶ 4.54.

[31]   *Id.* at ¶ 4.93.

[32]   The District Court had declined to hear the claims under the ECHR on procedural grounds. The Hague Court of Appeals and Supreme Court concluded that the claims were properly before the court because although Article 34 of the ECHR does not allow public interest actions to be brought before the European Court of Human Rights, that limitation did not apply to Dutch courts and that Urgenda could bring an action under section 305a of Dutch Civil Code, which allows class actions of interest groups.

[33]   Urgenda, Supreme Court decision ¶ 5.2.2.

protect individuals against possible serious damage to their environment," including where "there is a risk that serious environmental contamination may affect individuals' well-being and prevent them from enjoying their homes in such a way as to affect their private and family life adversely."[34] The Supreme Court held that "the Netherlands is obliged to do 'its part' in order to prevent dangerous climate change" because there is a grave risk that dangerous climate change will occur that will endanger the lives and welfare of many people in the Netherlands.[35] In ascertaining what it means for the Netherlands to do "its part," the court undertook a close and sophisticated analysis of climate projections and the international climate treaty regime. The court found that "there is a high degree of international consensus on the urgent need for the Annex I countries to reduce greenhouse emissions by at least 25–40% by 2020 compared to 1990 levels, in order to achieve at least the two-degree target" and that "[t]his high degree of consensus can be regarded as common ground within the meaning of the ECtHR case law."[36] The court noted the existence of an emissions gap, explained that "every reduction means that more room remains in the carbon budget,"[37] and recognized that "each postponement of a reduction in greenhouse gas emissions will require a future reduction to be more stringent in order to stay within the confines of the remaining carbon budget."[38] The court was thus rigorous and independent in defining the emissions reductions required to meet the duty of care and evaluating whether the Netherlands had satisfied that duty. At the same time, the court rejected arguments that it was overstepping its bounds by entering into the political domain, explaining that the government and parliament possess "a large degree of discretion to make the political considerations that are necessary in this regard," but that it is nonetheless "up to the courts to decide whether, in availing themselves of this discretion, the government and parliament have remained within the limits of the law by which they are bound," here Articles 2 and 8 of the ECHR.[39]

B   Greenpeace Nordic Association v. Ministry of Petroleum and Energy

In 2016, the Norwegian government awarded ten exploration and production licenses to energy companies under the Norwegian Petroleum Act giving them exclusive rights to conduct surveys and search for and produce petroleum in

---

[34]   Urgenda, Supreme Court decision ¶ 5.2.3.
[35]   Urgenda, Supreme Court decision ¶ 5.7.1.
[36]   Urgenda, Supreme Court decision ¶ 7.2.11.
[37]   Urgenda, Supreme Court decision ¶ 5.7.8.
[38]   Urgenda, Supreme Court decision ¶ 4.6.
[39]   Urgenda, Supreme Court decision ¶ 8.3.2.

specified areas of the Barents Sea South and South-East.[40] Prior to the issuance of the licenses, the areas had been "opened" for exploration under the Act, which requires that the Norwegian Ministry of Petroleum and Energy conduct an impact assessment and obtain approval of the Norwegian parliament, the Storting. The relevant areas in the Barents Sea South were opened for exploration in 1989 (and already have active areas in production); the areas in the Barents Sea South-East were opened for petroleum activities in 2013, the first opening of a new area in the Barents Sea in 24 years.

Environmental groups Greenpeace and Nature & Youth filed suit seeking to invalidate the licensing decision on the grounds that it violated the Norwegian Constitution (primarily Article 112) and Articles 2 and 8 of the ECHR and failed to satisfy relevant procedural requirements. The Oslo District Court, the Borgarting Court of Appeal, and the Norwegian Supreme Court all ruled to uphold the licensing decision, although the reasoning of the Court of Appeal and Norwegian Supreme Court differ significantly with respect to the scope and meaning of Article 112.[41] Article 112 of the Norwegian Constitution provides:

> Every person has the right to an environment that is conducive to health and to a natural environment whose productivity and diversity are maintained. Natural resources shall be managed on the basis of comprehensive long-term considerations which will safeguard this right for future generations as well. In order to safeguard their right in accordance with the foregoing paragraph, citizens are entitled to information on the state of the natural environment and on the effects of any encroachment on nature that is planned or carried out. The authorities of the state shall take measures for the implementation of these principles.[42]

Norway revised Article 112 in 2014 and many important questions about the meaning and enforcement of Article 112 (generally, aside from its application to climate change) presented matters of first impression.

---

[40] The ten exploration and production licenses authorize the search for oil; upon discovering oil, a process also governs extraction and requires submission and approval of a plan for development and operation.

[41] Nature & Youth and Greenpeace Nordic v. Norway, no. 16-166674TVI-OTIR/06. Unofficial English translations of the District Court and Court of Appeals decision [hereinafter Greenpeace Nordic Ass'n, Court of Appeal decision] and the Supreme Court's reading of its decision [hereinafter Greenpeace Nordic Ass'n Supreme Court decision] are available from CLIMATE CASE CHART, Greenpeace Nordic Ass'n v. Ministry of Petroleum and Energy, http://climatecasechart.com/non-us-case/greenpeace-nordic-assn-and-nature-youth-v-norway-ministry-of-petroleum-and-energy [https://perma.cc/WL84-D5VP] (last visited Apr. 8, 2021).

[42] KONGERIKET NORGES GRUNNLOV [CONSTITUTION], May 17, 1814, art. 112 (Nor.).

Despite ruling to uphold the challenged licensing decisions, in applying Article 112, the Court of Appeal made a number of threshold determinations regarding the application of Article 112 that suggested Article 112 might have robust application to climate change. The Court of Appeal concluded that Article 112 grants substantive rights that can be reviewed before the courts. It also reasoned that although the "key" when evaluating a decision under Article 112 is "effects arising in Norway," the court allowed that "it could nevertheless be a relevant element regarding actions based on Norway that also contribute to environmental harm outside Norway."[43] The court further concluded that international agreements "contribute to clarifying what is an acceptable tolerance limits and appropriate measures" and "could therefore be an important element in the overall assessment" of whether government conduct complies with Article 112.[44] With respect to how Article 112 applies to climate change specifically, the Court of Appeal held that while "individual emissions cannot be assessed in isolation, the assessment must nevertheless be based on the emissions that will result from the decision being challenged" and that in evaluating whether the government has violated Article 112, it is necessary to consider "local environmental harm, greenhouse gas emissions that occur in connection with the production of petroleum *and* greenhouse gas emissions that occur in connection with combustion."[45] Although the Court of Appeal ultimately upheld the challenged licensing decisions after applying this framework,[46] its recognition that Article 112 grants rights that can be asserted

---

[43]   Greenpeace Nordic Ass'n, Court of Appeal decision, p. 22.
[44]   Greenpeace Nordic Ass'n, Court of Appeal decision, p. 22.
[45]   Greenpeace Nordic Ass'n, Court of Appeal decision, p. 11, 21 (emphasis added).
[46]   With respect to emissions within Norway associated with the production process, the Court of Appeal noted that the process would produce a relatively small volume of emissions and, while recognizing that the Paris Agreement's targets and burden-sharing principles are "directly in opposition to searching for new discoveries," ultimately held that the Norwegian government possesses the flexibility and authority to allow production emissions associated with the granted licenses and choose to reduce other Norwegian emissions. Greenpeace Nordic Ass'n, Court of Appeal decision, p. 28 ("The prioritisations in climate policy involve socio-economic and political balancing in the core area for what the courts should be constrained in reviewing."). With respect to the more significant volume of emissions associated with combustion of produced and exported oil, the Court of Appeal emphasized the uncertainty of whether ceasing Norwegian oil exports would actually reduce worldwide GHG emissions, reasoned that climate change harms within Norway were "key" and likely to be less severe, noted that "the matter involves socio-economic and political balancing on which positions are continuously being taken in the Storting and which is in the core area for what the courts should be constrained in reviewing," and looked to the fact that the international climate treaty regime does not assign responsibility for emissions to exporting countries. Greenpeace Nordic Ass'n, Court of Appeal decision, p. 31.

before the courts and its willingness to consider international agreements and extraterritorial GHG emissions in evaluating government compliance with Article 112 suggested the possibility that Article 112 could shape government obligations regarding climate change.

The Supreme Court of Norway, however, adopted a narrower understanding of Article 112 and its application in the context of climate change that will likely prevent, or at minimum hinder, the use of Article 112 to require enhanced climate change mitigation policy in Norway.[47] The Supreme Court held that under Article 112 "a right can only be directly based on the constitutional provision to a limited degree in a case before the courts,"[48] explaining that:

> Article 112 of the Constitution must be read, when the Storting has considered a matter, as a safety valve. In order for the courts to set aside a legislative decision by the Storting, the Storting must have grossly disregarded its duties under the third paragraph of Article 112. The same must apply for other Storting decisions and decisions to which the Storting has consented. The threshold is consequently very high.[49]

The Supreme Court appears to have understood Article 112 to create "a standard for judicial review only in cases concerning environmental problems in which the legislator has not taken a position on the environmental problem at stake" and to "only set[ ] a material limit for state action and establish[ ] the standard for judicial review in situations where the duty under Article 112, paragraph 3, was grossly neglected," thereby "establish[ing] a very high threshold for setting aside legislative and other decisions that the parliament has taken or has consented to."[50] The decision means that courts in Norway will, in all but the most exceptional circumstances, defer to political (legislative) judgments about Norway's duty to mitigate climate change, including what is required by Article 112 in this regard.[51] The Supreme Court also held that Article 112 is

---

[47]   For a nuanced and thorough review of the decision by a Norwegian legal scholar, *see* Christina Voigt, *The Climate Judgment of the Norwegian Supreme Court: Aligning the Law with Politics* (February 23, 2021), https://ssrn.com/abstract=3791366 or http://dx.doi.org/10.2139/ssrn.3791366.

[48]   Greenpeace Nordic Ass'n Supreme Court decision ¶ 144.

[49]   *Id.* ¶ 142.

[50]   Voigt, *supra* note 47, at 8.

[51]   Of note, the Supreme Court did find that statutes and regulations, such as the Norwegian Petroleum Act and the Norwegian Petroleum Regulations, "must be interpreted and applied in light of Article 112 of the Constitution" and that "administrative proceedings, when opening new areas, must thoroughly clarify the advantages and disadvantages of the opening." Greenpeace Nordic Ass'n Supreme Court decision ¶ 184. Thus, Article 112 may promote climate-friendly interpretation and enforcement of

limited to environmental harms that manifest in Norway, explaining that extra-territorial combustion and emissions are relevant to Article 112 only to the extent that they increase harms in Norway.[52] And, in interpreting Article 112, the Supreme Court considered developments and discussions prior to or contemporaneous with the adoption of Article 112; it did not incorporate subsequent developments or international agreements, such as the Paris Agreement.

The Supreme Court of Norway also held that the connection between the emissions that would be generated by the issuance of the licenses and climate change impacts on Norwegians was too attenuated to present a "real and immediate risk" of loss of life so as to violate Article 2 of the ECHR[53] or to constitute a direct and immediate relationship between the environmental impairment and private life, family life or the home as required to establish a violation of Article 8 of the ECHR.[54] The court also reviewed and dismissed various procedural errors alleged by plaintiffs grounded in a failure to adequately assess impacts, including a number related to climate change.

C   Juliana v. United States

Juliana v. United States (*Juliana*) is the most significant of a group of climate change cases brought under the public trust doctrine. The nongovernmental organization Our Children's Trust supports many of the actions which, while brought in different jurisdictions and tailored to the law of those jurisdictions, broadly seek to compel stronger government action to mitigate climate change by invoking the public trust doctrine. The public trust doctrine recognizes a government duty to manage and protect public natural resources such as navigable waters, fisheries, and wildlife. By exacerbating or failing to mitigate climate change, the argument goes, governments violate their duty to preserve public natural resources that constitute the *res* of the public trust. Many states recognize the public trust doctrine in common law and some states have enshrined trust principles within their constitutions. There is, however, consid-

---

existing laws and regulations despite its limited ability to more directly influence the ambition of Norwegian climate change mitigation policy.

[52]   Greenpeace Nordic Ass'n Supreme Court decision ¶ 149 ("Article 112 of the Constitution does not generally protect against acts and effects outside the Kingdom of Norway. But if activities abroad that Norwegian authorities have directly influenced or could take measures against cause harm in Norway, this must be capable of being included through the use of Article 112. One example is combustion abroad of oil or gas produced in Norway, when it leads to harm in Norway as well.").

[53]   Greenpeace Nordic Ass'n Supreme Court decision ¶ 168 ("Firstly, it is uncertain whether or to what degree the decision actually will lead to emissions of greenhouse gases. Secondly, the possible effect for the climate is a good piece into the future.").

[54]   Greenpeace Nordic Ass'n Supreme Court decision ¶ 171.

erable variation between states with respect to which resources they recognize as falling within the *res* of the public trust and the requirements and implementation of the doctrine. On the federal level, there remains uncertainty about whether a federal public trust doctrine exists and binds the federal government and, if it does, the scope of its requirements.

As of this writing, no court in the US has required government climate mitigation under the public trust doctrine. However, before ultimately being dismissed by the Ninth Circuit Court of Appeals on other grounds, the public trust claim in *Juliana* survived motions to dismiss and for summary judgment, with a district court finding that the public trust doctrine, in conjunction with the Due Process Clause, can be understood to compel the government to mitigate climate change. In *Juliana*, a group of youth, a youth environmental advocacy group (EarthGuardians), and Dr. James Hansen (as guardian on behalf of future generations) sought an order requiring the federal government to act to reduce emissions. They alleged that a wide range of affirmative actions by the US government facilitating and promoting the production and combustion of fossil fuels contribute to climate change, thereby contravening the government's duties under the public trust doctrine and violating plaintiffs' rights under the Constitution (primarily the Due Process Clause). The district court, in its lengthy decision denying a motion to dismiss, found that the public trust doctrine constrains the federal government, encompasses within its *res* territorial seas being harmed by climate change, and is not subject to displacement by federal statutes (because the doctrine imposes a constitutionally protected limitation on the government). The district court further reasoned that the Due Process Clause provided the cause of action for plaintiffs to raise their public trust claim, determining that "plaintiffs' public trust claims are properly categorized as substantive due process claims."[55] Substantive due process prohibits government action that infringes a fundamental right unless the action is narrowly tailored to advancing a compelling government interest—a strict scrutiny standard which is rarely met. The district court held that the plaintiffs had adequately alleged infringement of a fundamental right because "the right to a climate system capable of sustaining human life is fundamental to a free and ordered society."[56] The district court thus merged the plaintiffs' public trust and constitutional claims and, adopting expansive interpretations of both bodies of law, found that plaintiffs' claim could proceed.

The district court declined to grant government motions to dismiss and for summary judgment and, after each of those decisions, also denied the govern-

---

[55]   Juliana v. United States, 217 F. Supp. 3d 1224, 1261 (D. Or. 2016), *rev'd and remanded*, 947 F.3d 1159 (9th Cir. 2020), 1261.

[56]   *Id.* at 1250.

ment's motions for certification of interlocutory appeal. The government took the unusual step of filing mandamus petitions seeking to effectively terminate further proceedings in the district court and sought stays during the pendency of those petitions. The Supreme Court denied the mandamus petitions, but in its decision not to grant a stay during the pendency of the first mandamus petition, it signaled its concerns about the case, observing that "the breadth of respondents' claims is striking ... and the justiciability of those claims presents substantial grounds for difference of opinion."[57] And in its decision denying defendants' second application for a stay—a decision drawing dissents from two justices, who would have granted the application for a stay—the Court seemed to nudge the Ninth Circuit to intervene, observing that many of the reasons that the Ninth Circuit has "twice denied the Government's request for relief" were "no longer pertinent."[58] The government never obtained a mandamus petition or stay, but the motions achieved the desired effect: The district court ultimately felt constrained to certify the case for interlocutory appeal, which the Ninth Circuit—over a dissent (lamenting that the district court appeared to have been pressured into certifying the appeal)—granted. As stated above, the Ninth Circuit then reversed the district court's denial of the government's motion for summary judgment.

These procedural details illustrate the anxiety that *Juliana* triggered about the potential for the judiciary to overstep its bounds with respect to climate change policy. This anxiety about the proper role of courts proved fatal to the case. On appeal, the Ninth Circuit dismissed the case, concluding that the plaintiffs lacked standing for reasons grounded in concerns about the propriety of judicial involvement in climate change policy. The Ninth Circuit held that the plaintiffs' claims were not redressable by the court because the "specific relief they seek is [not] within the power of an Article III court" as "it is beyond the power of an Article III court to order, design, supervise, or implement the plaintiffs' requested remedial plan."[59] Despite the plaintiffs' insistence that the court could order the development of a remedial plan while leaving its details (and associated policy decisions) to be made by the executive and legislative branches (a perspective that the Dutch courts found persuasive in a similar context in *Urgenda*), the Ninth Circuit found that this approach would ultimately "require the judiciary to pass judgment on the sufficiency of the government's response to the order, which necessarily would entail a broad range of policymaking."[60]

---

[57]   United States v. U.S. Dist. Ct. for Dist. of Oregon, 139 S. Ct. 1 (2018).

[58]   In re U.S., 139 S. Ct. 452, 453 (2018).

[59]   Juliana v. United States, 947 F.3d 1159, 1171 (9th Cir. 2020).

[60]   *Id.* at 1172.

III   THEMES, PERSPECTIVES, AND FUTURE DIRECTIONS

Together, the decisions described above reveal how actions brought to compel government mitigation challenge courts and highlight the difficulties facing such suits going forward while also suggesting the potential for these actions to strengthen government mitigation efforts.[61] Plaintiffs can readily demonstrate (and courts seem quite willing to accept) that climate change demands swift government action to avoid catastrophic consequences. Cases asking the judiciary to respond to this reality by requiring government mitigation action challenge courts in two important respects.

First, as the drafters of the Oslo Principles observe, there is no direct precedent or clear command for government action under a single body of law. Identifying and defining a government duty to mitigate requires weaving together myriad "intersecting sources" of law in novel ways. Recall that the lower court in *Urgenda* identified the Dutch Civil Code of hazardous negligence as the source of government duty to mitigate climate change while the intermediate appellate and Supreme Court grounded the state's duty in human rights law; all three courts cited extensively to a wide range of sources of law in defining the scope of the state's duty. In *Greenpeace Nordic Association*, the Court of Appeal noted that the Paris Agreement leaves governments with the discretion to choose how to meet emissions targets (and to rely on emissions trading or other flexibility mechanisms to do so, even if it means that they can "buy" themselves out of national emission cuts). In the court's view, this indicated that even though "searching for new discoveries" is "directly in opposition" to the Paris Agreement's burden-sharing principles and requirement for the highest possible ambition as applied to Norway, that tension resides within the Paris Agreement itself. And the Supreme Court of Norway cited to the practice of locating emissions accounting under international

---

[61]   These cases provide illustrative examples; numerous other cases raise similar claims. *E.g.*, Commune de Grande-Synthe v. France (alleging, in a suit brought by a municipality, that the French government's failure to take further action to reduce greenhouse gas emissions violates domestic and international law); Friends of the Irish Environment v. Ireland (holding that the Irish government's approval of the National Mitigation Plan in 2017 violated Ireland's Climate Action and Low Carbon Development Act 2015 but finding plaintiffs lacked standing to raise claims under the constitution or ECHR); Leghari v. Pakistan (holding that the Pakistani government's lassitude in implementing relevant climate laws violated citizens' fundamental rights and ordering specific government action); Union of Swiss Senior Women for Climate Protection v. Swiss Federal Council) (dismissing a claim brought by a group of senior Swiss women alleging that the Swiss government's inaction on climate change violated multiple provisions of the Swiss Constitution and the ECHR).

climate agreements to support its conclusion that issuance of the production licenses did not constitute a gross disregard of duty under Article 112, observing that "the Storting and the Government base Norwegian climate policy on the division of responsibilities that results from international agreements. A clear principle applies here that each state is responsible for the combustion that occurs on its own territory."[62] In *Juliana*, the district court wove together (in an unprecedented fashion) the plaintiffs' public trust doctrine claim and their constitutional claims to recognize a cause of action undergirded by the public trust doctrine but operationalized through the Due Process Clause.

Second, courts are asked to combine intersecting sources of law in novel ways in a context where they are deeply uneasy about overstepping the bounds of the judiciary. While courts are readily convinced of the factual imperative to mitigate climate change, they are often inclined to understand mitigation policy as the province of executives and legislatures, viewing those branches of government as better equipped (as a matter of institutional competence) and more clearly empowered (as a matter of constitutional law) to develop mitigation policy. The court in *Urgenda* addressed objections of judicial overreach, explaining that the court's order left discretion to the government to decide what actions to take to reduce emissions. The court noted that the "order does not amount to an order to take specific legislative measures, but leaves the State free to choose the measures to be taken in order to achieve a 25% reduction in greenhouse gas emissions by 2020."[63] The court then reasoned that, while "in the Dutch constitutional system of decision-making on the reduction of greenhouse gas emissions is a power of the government and parliament" over which those entities "have a large degree of discretion to make the political considerations that are necessary," nonetheless "[i]t is up to the courts to decide whether, in availing themselves of this discretion, the government and parliament have remained within the limits of the law by which they are bound."[64] Meanwhile, in *Greenpeace Nordic Association*, judicial restraint grounded in deference to legislative prerogatives was a significant factor contributing to the courts' conclusions that the threshold for finding a violation of Article 112 was not met. As explained by the Court of Appeal, the threshold for finding a violation was high because "the matter involves socio-economic and political balancing on which positions are continuously being taken in the Storting and which is in the core area for what the courts should be constrained in reviewing."[65] And although the plaintiffs argued (similar to the court's view

[62]   Greenpeace Nordic Ass'n, Supreme Court decision ¶ 159.
[63]   Urgenda, Supreme Court decision ¶ 8.2.7.
[64]   Urgenda, Supreme Court decision ¶ 8.3.2.
[65]   Greenpeace Nordic Ass'n, Court of Appeal decision p. 31.

in *Urgenda*) that the court could avoid encroaching on legislative prerogatives by leaving the details of mitigation to the government, the Ninth Circuit dismissed the *Juliana* case squarely for the reason that, in its view, the relief sought exceeded the scope of judicial authority, remarking that "any effective [remedial] plan would necessarily require a host of complex policy decisions entrusted, for better or worse, to the wisdom and discretion of the executive and legislative branches."[66]

What does this augur for actions to force government mitigation action going forward? Perhaps future cases will build on these early actions and spur the incremental growth of a body of law under which courts impose minimum requirements for government mitigation action that are grounded in sources of law (constitutions, human rights law, the common law) less likely to depend upon or be imperiled by lack of popular or political will. A follow-up suit to *Juliana*, for example, might build on the district court's recognition of a public trust theory suit brought under substantive due process in a case requesting more limited and specific relief to avoid presenting the redressability problems identified by the Ninth Circuit. As a normative matter, judicial engagement to ensure strong government efforts to mitigate climate change seems prudent even if jurisdictions adopt strong domestic laws requiring emission reductions. Judicially enforced minimum obligations for government mitigation grounded in more durable sources of law less susceptible to the vicissitudes of current political desires, like constitutions, human rights law, and the common law, may prove crucial to secure robust and consistent implementation of mitigation laws and to strengthen those laws as necessary to achieve needed emissions reductions.

CONCLUSION

This chapter reveals a stark difference in the posture courts adopt toward government with respect to climate change mitigation in different contexts. When courts understand themselves to be interpreting and enforcing mitigation commitments to which governments can be understood to have bound themselves through regular democratic processes (as with respect to use of statutes of general application to engage in climate change mitigation), courts appear quite comfortable defining the scope of government obligations to mitigate climate change. When, however, courts understand themselves to be in the position of imposing mitigation responsibilities upon governments based on authorities external to traditional democratic lawmaking processes (for example, grounded in the common law or constitutions), courts evince

---

[66] Juliana v. United States, 947 F.3d 1159, 1171 (9th Cir. 2020).

deep unease and show less willingness to order government mitigation action. There are certainly exceptions (as in the *Urgenda* case), but this generalization seems fair. Broadly speaking, plaintiffs asking courts to compel governments to adopt more robust mitigation law and policy, not anchored in existing domestic law, confront a strong undercurrent of judicial restraint. Cases to compel governments to honor commitments expressed in existing domestic legislation face less judicial reticence; once climate law exists—in the form of a statute of general application or climate-specific legislation—courts are quite comfortable insisting upon its robust implementation.

6. Human rights and climate change

Smita Narula

The impact of climate change on a range of human rights is now well documented, as are the disproportionate harms suffered by marginalized groups and vulnerable communities. In turn, it is increasingly well understood that states' fulfillment of their human rights obligations can lend vital support to their efforts to mitigate and adapt to climate change. This chapter examines the relationship between human rights and climate change using the framework of international human rights law. Part I provides an overview of the foundations, principles, and mechanisms of human rights law. Part II looks at the impact of climate change on a range of human rights. Part III considers how human rights law has evolved to elaborate states' obligations to mitigate and adapt to climate change, and to address the needs of those most vulnerable to climate-related harms. Part IV then analyzes the opportunities and challenges of using international human rights law as a legal framework to address the climate crisis.

I   THE FOUNDATIONS, PRINCIPLES, AND MECHANISMS OF INTERNATIONAL HUMAN RIGHTS LAW

The genesis of the modern international human rights system is often traced to the post-World War II prosecution of Nazi war criminals in the Nuremberg trials and the international community's collective desire to "prevent the recurrence of such crimes against humanity through development of new standards for the protection of human rights."[1] These standards were subsequently codified in four stages: the articulation of human rights concerns in the United Nations (UN) Charter;[2] the identification of specific rights in the Universal

---

[1]   Louis B. Sohn, *The New International Law: Protection of the Rights of Individuals Rather than States*, 32 Am U.M. Rev. 1, 10 (1982).

[2]   U.N. Charter art. 1 (purpose of the United Nations is to achieve international cooperation to solve economic, social, cultural and humanitarian problems while promoting human rights for all without distinction); *id.* art. 13 (role of General Assembly is

Declaration of Human Rights (UDHR);[3] the elaboration of each of the rights in the International Covenant on Civil and Political Rights (ICCPR)[4] and the International Covenant on Economic, Social and Cultural Rights (ICESCR);[5] and the adoption of additional conventions and declarations concerning various human rights issues, including gender and racial discrimination, torture and forced disappearance, and the rights of children, migrants, and persons with disabilities.[6] Additional instruments have been adopted at the regional level. Human rights protections are also reflected in the constitutions and other laws of most nations.

By ratifying and becoming parties to international human rights treaties, states commit to putting in place domestic measures and legislation to respect, protect, and fulfill human rights. The obligation to *respect* human rights means that states must refrain from acting in a way that would interfere with or curtail people's ability to exercise or enjoy their human rights. The obligation to *protect* requires states to protect individuals and groups from human rights abuses, including abuses committed by third party actors such as corporations. And the obligation to *fulfill* human rights means that states must take proactive steps to facilitate the enjoyment of human rights.[7]

---

to study and make recommendations to promote international cooperation and the realization of human rights); *id.* art. 55 (U.N. shall promote respect for human rights).

[3]   Adopted in 1948, the UDHR is a foundational document of international human rights law that set out, for the first time, fundamental human rights that must be universally protected.

[4]   International Covenant on Civil and Political Rights, Dec. 16, 1966, 999 U.N.T.S. 171 [hereinafter ICCPR].

[5]   International Covenant on Economic, Social and Cultural Rights, Dec. 16, 1966, 993 U.N.T.S. 3 [hereinafter ICESCR].

[6]   *See, e.g.*, Convention on the Rights of the Child, G.A. Res. 44/25, annex, U.N. GAOR, 44th Sess., Supp. 49 at 167, U.N. Doc. A/44/49 (Nov. 20, 1989) [hereinafter CRC]; Convention on the Elimination of All Forms of Discrimination Against Women, U.N. GAOR, 34th Sess., G.A. Res. 341180, Supp. No. 46 at 193, U.N. Doc. A/34/46 (Dec. 18, 1979) [hereinafter CEDAW]; International Convention on the Elimination of All Forms of Racial Discrimination, Mar. 7,1966, 660 U.N.T.S. 195.

[7]   This three-level typology of states' duties—which was originally developed by Asbjørn Eide in his study on the right to adequate food as a human right—is now a widely used framework for analyzing states' human rights obligations generally. U.N. Econ. & Soc. Council, Sub-Comm'n on Prevention of Discrimination & Prot. of Minorities, *The New International Economic Order and the Promotion of Human Rights: Rep. on the Right to Adequate Food as a Human Right Submitted by Mr. Asbjørn Eide, Special Rapporteur,* ¶¶ 112–114, U.N. Doc. E/CN.4/Sub.2/1987/23 (July 7, 1978).

Governments must ensure human rights in a non-discriminatory manner,[8] and must take special care to ensure the rights of vulnerable groups, including those most vulnerable to environmental harms.[9] Where rights have been violated, states must also ensure an effective remedy.[10] When domestic proceedings fail to provide redress for human rights abuses, individuals can, under certain circumstances, submit complaints and communications at the regional and international level to help ensure that states live up to their human rights obligations.

At the international level, states' implementation of their human rights obligations are monitored by both treaty and UN Charter-based bodies. Treaty monitoring bodies—which are comprised of independent experts—evaluate states' performance under specific human rights treaties, carrying out a number of functions as prescribed by the provisions of the treaties under which they are established. Treaty monitoring bodies review periodic reports submitted by states' parties and offer recommendations for action. They also adopt "general comments" (sometimes called "general recommendations") reflecting their interpretation of states' obligations under specific treaty articles. Under certain conditions, a sub-set of treaty monitoring bodies may also conduct country inquiries and consider complaints by individuals who claim that their rights under the treaty have been violated.[11]

---

[8] *See, for example,* ICESCR, *supra* note 5, art. 2(2) ("The States Parties to the present Covenant undertake to guarantee that the rights enunciated in the present Covenant will be exercised without discrimination of any kind as to race, colour, sex, language, religion, political or other opinion, national or social origin, property, birth or other status."); and ICCPR, *supra* note 4, art. 2(1) ("Each State Party to the present Covenant undertakes to respect and to ensure to all individuals within its territory and subject to its jurisdiction the rights recognized in the present Covenant, without distinction of any kind, such as race, colour, sex, language, religion, political or other opinion, national or social origin, property, birth or other status.").

[9] John Knox (Special Rapporteur on the Issue of Human Rights Obligations Relating to the Enjoyment of a Safe, Clean, Healthy and Sustainable Environment), *Human Rights Obligations Relating to the Enjoyment of a Safe, Clean, Healthy and Sustainable Environment,* ¶ 22, U.N. Doc. A/73/188 (July 19, 2018) [hereinafter Mapping Report]; *see also The impacts of climate change on the effective enjoyment of human rights,* U.N. HUM. RTS. OFF. OF THE HIGH COMM'R, https://www.ohchr.org/EN/Issues/HRAndClimateChange/Pages/AboutClimateChangeHR.aspx (last visited Feb. 28, 2021) (noting that "States have a human rights obligation to prevent the foreseeable adverse effects of climate change and ensure that those affected by it, particularly those in vulnerable situations, have access to effective remedies and means of adaptation to enjoy lives of human dignity").

[10] Mapping Report, *supra* note 9, ¶ 14.

[11] For more information on the work of treaty monitoring bodies, see *Monitoring the core international human rights treaties,* U.N. HUM. RTS. COUNCIL, https://www.ohchr.org/EN/HRBodies/Pages/WhatTBDo.aspx (last visited Mar. 28, 2021).

The UN Human Rights Council—a Charter-based body—is an inter-governmental body comprising 47 states that advances the promotion and protection of all human rights around the globe.[12] A key function of the Human Rights Council is to conduct a Universal Periodic Review (UPR) of the human rights records of all UN Member States and recommend steps that each state can take to further fulfill its human rights obligations. Civil society groups can inform the UPR process by submitting information to be added to a "other stakeholders" report, which is considered alongside information provided by the state under review.[13]

The Human Rights Council also adopts resolutions and appoints independent human rights experts (often called Special Rapporteurs) with mandates to advise and report on human rights from either a thematic or country-specific perspective. Special Rapporteurs undertake country visits, send "communications" to states to act on individual cases of reported violations or on issues of broader concern, and conduct annual thematic studies that contribute to the development of international human rights standards and to a greater understanding of how specific rights may be violated.[14] Established by the UN General Assembly in 1993, the Geneva-based Office of the High Commissioner for Human Rights (OHCHR)—headed by the UN High Commissioner for Human Rights—is mandated to promote and protect all human rights and helps provide logistical, administrative, and substantive support to the human rights mechanisms described above.[15]

At the regional level, human rights systems have developed to reflect regional values and concerns and to allow for additional mechanisms for enforcing human rights standards. The regional systems in Europe, the Americas, and Africa are the most developed to date.[16] Based in Strasbourg,

---

[12]   The Human Rights Council holds three regular sessions a year at the UN Office in Geneva and can also hold special sessions to address human rights violations and emergencies. For more on the Human Rights Council, see *Welcome to the Human Rights Council*, U.N. Hum. Rts. Council, https://www.ohchr.org/EN/HRBodies/HRC/Pages/AboutCouncil.aspx (last visited Mar. 25, 2021).

[13]   For more on the UPR, see *Basic facts about the UPR*, U.N. Hum. Rts. Council, https://www.ohchr.org/EN/HRBodies/UPR/Pages/BasicFacts.aspx (last visited Mar. 28, 2021).

[14]   For more on the Special Procedures, see *Special Procedures of the Human Rights Council*, U.N. Hum. Rts. Council, ohchr.org/EN/HRBodies/SP/Pages/Introduction.aspx (last visited Mar. 28, 2021).

[15]   For more on the OHCHR, see *Who we are: an overview*, U.N. Hum. Rts. Off. of the High Comm'r, https://www.ohchr.org/EN/AboutUs/ (last visited Mar. 30, 2021).

[16]   The Asian and Arab regional systems are still developing. For more see *Asia*, International Justice Resource Center, https://ijrcenter.org/regional/asia/ (last visited May 5, 2021); *Middle East and North Africa*, International Justice Resource

France, and set up in 1959, the European Court of Human Rights (ECtHR) ensures that Member States of the Council of Europe uphold the rights and guarantees set out in the European Convention on Human Rights (ECHR).[17] The Court issues advisory opinions and examines individual complaints or "applications" alleging violations of human rights by a Member State.[18] The Court's judgments are binding and countries concerned must comply with them.[19] As the oldest regional court, the jurisprudence of the ECtHR is the most developed.

The inter-American system for the protection of human rights is responsible for monitoring and protecting human rights in the 35 countries that comprise the Organization of American States (OAS). This regional system is composed of two main entities: the Inter-American Commission on Human Rights (IACHR), based in Washington, D.C., and the Inter-American Court of Human Rights (IACtHR), based in San José, Costa Rica. The IACHR can receive individual complaints concerning alleged violations of the American Convention on Human Rights, while the IACtHR can receive complaints from the IACHR and from states' parties to the American Convention who have accepted the Court's jurisdiction.[20] The judgments of the Court are binding although the system lacks a means of enforcement. At the request of the Commission or of Member States, the Court may also issue advisory opinions concerning the interpretation of inter-American instruments. The Commission also engages in a range of human rights monitoring activities through the work of its various special rapporteurs.[21]

The youngest of the three regional systems, the African human rights system was created under the auspices of the African Union (AU). It too comprises a commission and a court with complementary mandates. The African

---

CENTER, https://ijrcenter.org/regional/middle-east-and-north-africa/ (last visited May 5, 2021).

[17] The ECHR, formally titled the "European Convention for the Protection of Human Rights and Fundamental Freedoms," was adopted in 1950 and came into force in 1953. Convention for the Protection of Human Rights and Fundamental Freedoms, Apr. 11, 1950, E.T.S. No. 005.

[18] Member States also sometimes submit complaints.

[19] For more on the Court, *see generally* EUR. CT. H.R., https://www.echr.coe.int/ (last visited Mar. 30, 2021).

[20] *See* Organization of American States, American Convention on Human Rights, Nov. 22, 1969, O.A.S.T.S. No. 36, 1144 U.N.T.S. 123, which was adopted in 1969 and entered into force in 1978. *See also* Inter-Am. Comm'n H.R., *Petition and Case System* 4, and 6 (2010), https://www.oas.org/en/iachr/docs/pdf/howto.pdf.

[21] For more on IACHR and IACtHR see *IACHR*, OAS, http://www.oas.org/en/ iachr/ (last visited Mar. 30, 2021); CORTE INTERAMERICANA DE DERECHOS HUMANOS, https://www.corteidh.or.cr/ (last visited Mar. 20, 2021).

Commission on Human and Peoples' Rights (ACHPR), now based in Banjul, The Gambia, and the African Court on Human and Peoples' Rights (ACtHPR), based in Arusha, Tanzania, both receive individual complaints and assess AU Member States' compliance with human rights standards under the African Charter on Human and Peoples' Rights,[22] though the activities and jurisdiction of each body are distinct.[23] As detailed in Parts II and III, the human rights bodies and mechanisms outlined above have contributed in significant ways to illuminating the human rights impacts of climate change, and elaborating on states' obligations therein.

II   THE HUMAN RIGHTS IMPACTS OF CLIMATE CHANGE

Increasing temperatures, rising sea levels, extreme weather events, melting permafrost and changes in precipitation patterns all have direct impacts on human rights, threatening the enjoyment of economic, social, and cultural rights, civil and political rights, and collective rights such as the right to self-determination. These impacts, which are now acutely felt around the globe, will intensify dramatically over the coming decades, impeding access to food, shelter, clean water, and other basic necessities for vast swathes of the world's population.

A   Economic, Social, and Cultural Rights

Economic, social, and cultural rights were initially codified in the ICESCR and have subsequently been enshrined in numerous constitutions. States that ratify the ICESCR make a commitment to progressively ensure the rights to an adequate standard of living (including adequate housing and food), health, education, family, and cultural life, as well as the right to social security, work, and good work conditions, among others.[24] The climate crisis deeply imperils a range of economic and social rights, including in particular the rights to health, food, water and sanitation, and housing.[25]

---

[22]   African (Banjul) Charter on Human and Peoples' Rights, OAU Doc. CAB/LEG/67/3, 21 I.L.M. 58 (June 27, 1981), which was adopted in 1981 and entered into force in 1986.

[23]   For more on the ACHPR and the ACtHPR, *see generally* Afr. Comm'n H.P.R., https://www.achpr.org/ (last visited Mar. 30, 2021); Afr. Ct. H.P.R., https://www.african-court.org/wpafc/ (last visited Mar. 30, 2021).

[24]   See ICESCR, *supra* note 5, arts. 6–15.

[25]   The rights to water and sanitation are not explicitly referenced in the ICESCR. In UN General Assembly resolution 64/292, states recognized the rights to water and

1   The right to health

An analytical study by the Office of the High Commissioner for Human Rights on the impact of climate change on the right to health noted that rising temperatures and more frequent and intense heatwaves will contribute to increases in heat-related deaths, especially in elderly populations.[26] Extreme weather events and natural disasters will continue to lead to injury, disability, death, and the transmission of infectious diseases.[27] Climate change can also affect nutrition through an increased loss of livelihood and poverty, changes in crop yields, and reduced access to food, water, and sanitation.[28] Moreover, climate change has profound impacts on mental health, which can deteriorate as a result of both the immediate and long-term physical effects of climate change.[29] As noted by the American Psychological Association, "Watching the slow and seemingly irrevocable impacts of climate change unfold, and worrying about the future for oneself, children, and later generations, may be an additional source of stress," a condition that is now referred to as "eco-anxiety."[30]

2   The right to food

Climate change patterns are expected to devastate agricultural production, while the dominant modes of food production and distribution are themselves contributing to environmental harms. In its special report on climate change and land, the Intergovernmental Panel on Climate Change (IPCC) concluded with "high confidence" that climate change had "already affected food security due to warming, changing precipitation patterns, and greater frequency of some extreme events."[31] In a report on climate change and the right to food, the former Special Rapporteur on the right to food similarly noted that rising tem-

---

sanitation as human rights, noting that those were essential for the full enjoyment of all human rights. The rights to water and sanitation are also referenced in the Convention on the Elimination of All Forms of Discrimination against Women.

[26]   Rep. of Off. of U.N. High Comm'r for Hum. Rts., *Analytical Study on the Relationship Between Climate Change and the Human Right of Everyone to the Enjoyment of the Highest Attainable Standard of Physical and Mental Health*, ¶ 12, UN Doc. A/HRC/32/23 (May 6, 2016) [hereinafter *Analytical Study on the Relationship between Climate Change and Human Rights*].

[27]   *Id.* ¶¶ 15, 17–19.

[28]   *Id.* ¶ 20.

[29]   *Id.* ¶ 21.

[30]   SUSAN CLAYTON ET AL., AM. PSYCH. ASS'N, CLIMATE FOR HEALTH & ECOAMERICA, MENTAL HEALTH AND OUR CHANGING CLIMATE: IMPACTS, IMPLICATIONS, AND GUIDANCE 27 (2017).

[31]   Intergovernmental Panel on Climate Change [IPCC], Summary for Policymakers, *Climate Change and Land: an IPCC special report on climate change, desertification, land degradation, sustainable land management, food security, and greenhouse gas fluxes in terrestrial ecosystems*, Finding A2.8 at 7 (2019).

peratures and extreme weather events will have significant global impacts on crop, livestock, fisheries, and aquaculture productivity, as will water scarcity and increasingly frequent droughts in arid regions.[32] "An increase of just 1°C in temperature," the report notes, "can have devastating impacts on crop yields and the ability to maintain current levels of agricultural production." Rising sea levels will also affect food availability in coastal areas and river deltas that are home to 60% of the world's population, which in turn will force greater climate-induced migration.[33]

3   The right to water

More than 2 billion people today live in countries experiencing high water stress (scarcity).[34] By 2050, almost twice as many are predicted to do so.[35] Climate change "makes water availability less predictable and increases the incidences of flooding that may destroy water points and sanitation facilities and contaminate water sources."[36] Both groundwater and surface water are also at risk from saltwater intrusion due to sea-level rise.[37] According to the IPCC, "climate change is projected to reduce renewable surface water and groundwater resources significantly in most dry subtropical regions," which will "intensify competition for water among agriculture, ecosystems, settlements, industry, and energy production, affecting regional water, energy, and food security."[38] This competition for scarce resources will in turn lead to greater social unrest and conflict.[39]

4   The right to housing

The right to adequate housing is threatened by climate change in numerous ways. Most starkly, extreme weather events can destroy homes and lead to displacement. Drought, erosion, and flooding can also render areas uninhabit-

---

[32]   U.N. Secretary-General, *Right to Food*, ¶¶ 7–8, U.N. Doc. A/70/287 (Aug. 5, 2015).

[33]   *Id.* ¶¶ 9–10.

[34]   U.N. Water, Sustainable Development Goal 6: Synthesis Report on Water and Sanitation 12 (2018).

[35]   U.N. Hum. Rts. Off. of High Comm'r, Frequently Asked Questions on Human Rights and Climate Change 13 (2021). [hereinafter OHCHR FAQ]

[36]   *Id.* at 12.

[37]   *Id.* at 13.

[38]   IPCC, *Part A: Global and Sectoral Aspects*, *in* Climate Change 2014: Impacts, Adaptation, and Vulnerability 232 (2014).

[39]   U.N. Secretary-General, *Question of the Realization in All Countries of Economic, Social and Cultural Rights*, ¶¶ 11, 15, U.N. Doc. A/HRC/37/30 (Dec. 18, 2017).

able in a more gradual manner, resulting in the same outcomes.[40] Those who are experiencing homelessness and those who lack resilient or secure housing will be the most adversely affected.[41] This may include LGBTQ+ individuals who may be more susceptible to homelessness and may lack access to stable housing as a result of being rejected by family members or facing discrimination in the workplace.[42] Rising temperatures can also affect the right to adequate housing, particularly in dense urban areas in what is known as the "heat island effect."[43]

5   The right to culture

The impact of the climate crisis on cultural rights was acutely described in a petition filed before IACHR. In 2005, Sheila Watt-Cloutier—an Inuk woman and Chair of the Inuit Circumpolar Conference—petitioned the IACHR to "obtain relief from human rights violations resulting from the impacts of global warming and climate change caused by acts and omissions of the United States."[44] On behalf of herself and 62 other named individuals, Watt-Cloutier requested that the Commission recommend that the US, *inter alia*, adopt mandatory measures to limit its GHG emissions and implement a plan to protect Inuit culture and resources.[45]

As articulated in the petition, "All aspects of the Inuit's lives depend upon their culture, and the continued viability of the culture depends in turn on the Inuit's reliance on the ice, snow, land and weather conditions in the Arctic."[46] The petition goes on to describe, across 33 pages and in painstaking detail, how changes in ice and snow conditions, thawing permafrost, coastal erosion, storm surges and flooding, changing species distribution, increasing temperatures and increasingly unpredictable weather in the Arctic have permanently damaged Inuit culture while undermining a range of other human rights.[47]

---

[40]   OHCHR FAQ, *supra* note 35, at 14.

[41]   *Id.*

[42]   Sophie Lee, *The Climate Crisis and the LGBTQ Community*, Sierra Club (June 25, 2019), https://www.sierraclub.org/washington/blog/2019/06/climate-crisis-and-lgbtq-community. The specific vulnerabilities of LGBTQ+ communities to climate change are often overlooked.

[43]   Heat islands are urban areas that experience higher temperatures than natural landscapes because structures such as roads and buildings absorb and re-emit the sun's heat.

[44]   Sheila Watt-Cloutier, et al. v. United States, Petition, Inter.-Am. Comm'r H.R. 1 (2005).

[45]   *Id.* at 118.

[46]   *Id.* at 35.

[47]   See *id.* at 35–67, which provides for plaintiffs' claims that "Global Warming Harms Every Aspect of Inuit Life and Culture."

Although the IACHR decided not to proceed with the petition, the case helped establish the critical linkage between human rights and climate change and also contributed to a 2007 IACHR thematic hearing on the subject.[48]

B   Civil and Political Rights

The civil and political rights articulated in the UDHR were later codified in the ICCPR, which obligates states' parties to respect and ensure, *inter alia*, the rights to life, liberty, and security of person; equality before the law; freedom of speech, assembly, and association; freedom of religion and privacy; freedom from torture, ill-treatment, and arbitrary detention; and the right to political participation.

1   The right to life
The ICCPR ensures the right to life as a fundamental and non-derogable right.[49] As noted by the Office of the High Commissioner for Human Rights, "[t]his entails at the very least, that States should take effective measures against foreseeable and preventable loss of life."[50] The UN Human Rights Committee, the treaty monitoring body which monitors states' compliance with their obligations under the ICCPR, has stated that "Environmental degradation, climate change and unsustainable development constitute some of the most pressing and serious threats to the ability of present and future generations to enjoy the right to life."[51]

2   Environmental defenders
Climate change can also affect the enjoyment of civil and political rights in indirect ways. For example, when climate-related policies (or policies that contribute to climate change) proceed without the meaningful participation of affected communities. Moreover, climate activists and land rights defenders who stand up to environmentally destructive industry practices are increasingly subject to violent attacks, arbitrary arrests, death threats, sexual violence, and

---

[48]   Human Rights and Global Warming: Hearing Before the Inter-Am. Comm'n H.R., 127th session, Mar. 1, 2007; *see also Inuit Petition and the IACHR, Campaign Update*, CTR. FOR INT'L ENV'T L., https://www.ciel.org/project-update/inuit-petition -and-the-iachr/ (last visited Mar. 29, 20201) (noting that that petition "helped broaden and re-focus the terms of the climate change debate").

[49]   See ICCPR, *supra* note 4, arts. 4, 6.

[50]   OFF. OF THE HIGH COMM'R FOR HUM. RTS., UNDERSTANDING HUMAN RIGHTS AND CLIMATE CHANGE 13 (2015) [hereinafter HUMAN RIGHTS AND CLIMATE CHANGE].

[51]   U.N. Hum. Rts. Comm., *General Comment No. 36, Article 6: right to life*, ¶ 62, U.N. Doc. CCPR/C/GC/36 (Sept. 3, 2019) [hereinafter *General Comment No. 36*].

spurious lawsuits.[52] Many are killed. According to the 2020 Global Witness report *Defending Tomorrow*, 2019 was the highest year on record for killings of land and environmental defenders, with 212 documented killings. More than half of the reported killings occurred in just two countries: Colombia and the Philippines.[53] As stated in the report:

> For years, land and environmental defenders have been the first line of defence against the causes and impacts of climate breakdown. Time after time, they have challenged the damaging aspects of industries rampaging unhampered through forests, wetlands, oceans and biodiversity hotspots. Yet despite clearer evidence than ever of the crucial role they play and the dangers they increasingly face, far too many businesses, financiers and governments fail to protect them in their vital and peaceful work.[54]

The NGO *Front Line Defenders*' global analysis of killings in 2020 found that at least 331 human right defenders were killed in 2020, 69% of whom were killed defending land, Indigenous peoples', or environmental rights.[55] These reports point to a disturbing and growing trend. In her December 2020 report to the UN Human Rights Council, the Special Rapporteur on the situation of human rights defenders presented data showing that since 2015, a total of 1,323 human rights defenders had been killed. The report notes that "Latin America is consistently the most affected region, and environmental human rights defenders are the most targeted."[56]

C   The Right to Self-Determination

Climate change also poses a serious threat to collective rights, such as the right to self-determination. Common article 1 of the ICESCR and the ICCPR states: "All peoples have the right to self-determination. By virtue of that right they freely determine their political status and freely pursue their economic, social and cultural development."[57] Climate change not only poses a significant threat to individuals' human rights and lives, "but also to their ways of life and live-

---

[52] GLOBAL WITNESS, DEFENDING TOMORROW 8 (2020).

[53] *Id.* at 4. The report notes that mining and extractives (oil and gas), agribusiness, and logging were the deadliest sectors. *Id.* at 9.

[54] *Id.*

[55] FRONT LINE DEFENDERS, FRONT LINE DEFENDERS: GLOBAL ANALYSIS 2020 4 (2020).

[56] See Mary Lawlor (Special Rapporteur on the Situation of Human Rights Defenders), *Final warning: death threats and killings of human rights defenders*, ¶¶ 5, 35–42, U.N. Doc. A/HRC/46/35 (Dec. 24, 2020).

[57] ICCPR, *supra* note 4, art. 1; ICESCR, *supra* note 5, art. 1.

lihoods, and to the survival of entire peoples."[58] This is perhaps most starkly seen with the dangers that sea-level rise and other climate-related harms pose to low-lying island states, for whom climate change represents an existential threat.[59] Climate change also threatens to deprive Indigenous peoples of their sources of livelihood and their traditional territories, undermining their right to self-determination.[60]

D   The Impact of Mitigation and Adaptation Measures on Human Rights

Climate change mitigation and adaptation strategies can themselves contribute to human rights harms. Mitigation projects undertaken to reduce or sequester GHG emissions, for example, can adversely affect the human rights of vulnerable groups. The most egregious examples involve large-scale land acquisitions (also known as "land grabs") for biofuels plantations and hydroelectric dam projects that can have profound environmental impacts,[61] can lead to forced evictions and mass displacement, and that more often than not proceed without meaningful consultation with or the participation of affected communities.[62] Under the UN Declaration on the Rights of Indigenous Peoples (UNDRIP), states must obtain Indigenous peoples' "free, prior and informed consent" before proceeding with projects that affect Indigenous territories[63]—a right

---

[58]   OHCHR FAQ, *supra* note 35, at 5.

[59]   *See* Hum. Rts. Council, *Annual Rep. Of the United Nations High Commission for Human Rights on the Relationship between Climate Change and Human Rights,* ¶ 40, U.N. Doc. A/HRC/10/61 (Jan. 15, 2009) [hereinafter *OHCHR 2009 Report*] (noting that climate change endangers the habitability and territorial existence of a number of low-lying island states).

[60]   OHCHR FAQ *supra* note 35, at 5–6.

[61]   *See* UNEP & COLUM. L. SCH. SABIN CTR. FOR CLIMATE CHANGE L., CLIMATE CHANGE AND HUMAN RIGHTS 8 (2015) [hereinafter UNEP, CLIMATE CHANGE AND HUMAN RIGHTS] (noting that hydroelectric projects can destroy the ecosystems on which communities depend and can "harm the health and livelihoods of people living downstream from the project by reducing river flows" and that biofuels projects "can contribute to food shortages and price shocks, additional water stress and scarcity, widespread deforestation, and displacement of indigenous peoples and small-scale farmers through land acquisitions").

[62]   For more on the human rights impacts of large-scale land deals, see Smita Narula, *The Global Land Rush: Markets, Rights, and the Politics of Food,* 49 Stan. J. Int'l L. 101, 109–121 (2013); DAVID K. DENG, ANDREA JOHANSSON & SMITA NARULA, N.Y.U. CTR. FOR HUM. RTS. & GLOB. JUSTICE, FOREIGN LAND DEALS AND HUMAN RIGHTS: CASE STUDIES ON AGRICULTURAL AND BIOFUEL INVESTMENT (2010).

[63]   G.A. Res. 61/295, U.N. Declaration on the Rights of Indigenous Peoples, art. 32 (Sept. 13, 2007). The principle of free, prior, and informed consent is also referenced throughout the Declaration.

that is often violated in the context of large-scale development projects.[64] Human rights can also be violated when states try to benefit from carbon emissions credits for clean development mechanism projects. For example, concerns have been raised about the potential effect of Reducing Emissions from Deforestation and Forest Degradation (REDD/REDD+) projects on Indigenous peoples and local communities whose input or consent may not be secured before proceeding with such projects and who may be displaced as a result.[65]

Measures taken to adapt to the impacts of climate change can also affect human rights. For example, human rights concerns can arise with the construction of coastal fortifications to adapt to sea-level rise, which may be erected without appropriate consultation with affected communities and which may protect one community at the expense of another.[66] Resettlement or relocation programs may similarly be carried out without adequate input or consent from those affected.[67]

Another issue now looming on the horizon involves the potential effects of geoengineering on human rights. Geoengineering refers to a set of emerging technologies that involve the deliberate and large-scale manipulation of the Earth's climate system through, for example, solar radiation management and ocean iron fertilization. Although field tests are still nascent and relatively small, according to studies cited by the UN Environment Programme (UNEP), such projects could "seriously interfere with the enjoyment of human rights for millions and perhaps billions of people" by significantly disrupting ocean and territorial ecosystems—which in turn could affect access to food, water, and other key resources—and by causing region-wide changes in precipitation.[68]

---

[64]   Christina Hill, Serena Lillywhite & Michael Simon, Oxfam Australia, Guide to Free Prior and Informed Consent 2 (2010).

[65]   UNEP, CLIMATE CHANGE AND HUMAN RIGHTS, *supra* note 61, at 9; *see also Analytical Study on the Relationship between Climate Change and Human Rights*, *supra* note 26, ¶ 5 (noting that marginalized communities are especially vulnerable to climate change and to mitigation and adaption responses to climate change, and providing as examples that the "biofuel agro-industry, hydroelectric power and forest conservation efforts can contribute to food insecurity and displacement").

[66]   Michael Burger and Jessica Wentz, *Climate Change and Human Rights, in* HUMAN RIGHTS AND THE ENVIRONMENT: LEGALITY, INDIVISIBILITY, DIGNITY AND GEOGRAPHY 202 (James May & Erin Daly eds., 2019); UNEP, CLIMATE CHANGE AND HUMAN RIGHTS, *supra* note 61, at 10.

[67]   UNEP, CLIMATE CHANGE AND HUMAN RIGHTS, *supra* note 61, at 10.

[68]   *Id.*

Unsettlingly, geoengineering projects are proceeding without any real oversight mechanisms in place.[69]

E   Unequal Burden and Unequal Blame

The human rights impacts of climate change have and will be most acutely
felt by countries in the Global South and by marginalized communities and
vulnerable groups across the Global North and South. This includes women,
children, older persons, Indigenous peoples, rural communities, persons living
in poverty, communities of color, national, ethnic, religious, or linguistic
minorities, caste-affected communities, migrants, persons with disabilities,
and those living in geographically vulnerable areas. Many individuals also
experience vulnerability along multiple axes.[70]

Indigenous peoples are particularly vulnerable to climate change because of
their proximity to and close relationship with natural ecosystems that are part
of their territories.[71] Indigenous peoples are affected both by the impacts of
climate change—including extreme weather events, drought, melting ice, and
sea-level rise—as well as the environmentally destructive activities that are
taking place on their territories and that are contributing to climate change—
such as deforestation, land grabs, and the exploitation of mineral resources.[72]
Peasants and rural communities face similar vulnerabilities, as was recognized
by the UN General Assembly with its adoption in 2018 of the UN Declaration
on the Rights of Peasants and Other People Working in Rural Areas.[73]

Children's physiology and developmental needs put them at particular risk
of the effects of climate change, leading the UN Committee on the Rights of
the Child—which monitors states' implementation of their obligations under

---

[69]   *Id.*; *see also* Railla Veronica D. Puno, *A Rights-Based Approach to Governance
of Climate Geoengineering*, 50 Env't L. Rep. 10744 (2020) (arguing for a rights-based
approach to the governance of geoengineering in international law that "would take into
account the need for participation, accountability, nondiscrimination, and equality in its
development and deployment, while addressing the potential of such technologies in
mitigating the impacts that climate change would have to the full enjoyment of human
rights, including the right to a healthy environment").

[70]   Mapping Report, *supra* note 9, ¶ 22.

[71]   See, for example, *Id.* ¶ 23, which provides a more general context of Indigenous
peoples' vulnerability to environmental harm; *see also* Rishabh Kumar Dhir et al.,
Int'l Labour Off., Indigenous Peoples and Climate Change: From Victims to
Change Agents through Decent Work 7–21 (2017) (documenting how Indigenous
peoples are affected by climate change in distinct ways and are also affected by policies
aimed at addressing climate change).

[72]   OHCHR FAQ, *supra* note 35, at 20.

[73]   G.A. Res. 39/12, U.N. Declaration on the Rights of Peasants and Other People
Working in Rural Areas, pmbl., (Sept. 28, 2018).

the Convention on the Rights of the Child—to describe climate change as one of the biggest threats to children's health.[74] The former Special Rapporteur on human rights and the environment stated that "climate change and the loss of biodiversity threaten to cause long-term effects that will blight children's lives for years to come,"[75] concluding that no group is more vulnerable to environmental harm than children.[76]

The gender-based impacts of climate change are also an increasing area of focus and concern for human rights bodies. In its General Recommendation No. 37 on the gender-related dimensions of disaster risk reduction in the context of climate change, for example, the Committee on the Elimination of Discrimination Against Women[77] noted that women and girls experience greater risks, burdens, and impacts of climate change and associated disasters than do men and boys.[78] And that's because crisis situations can serve to "exacerbate pre-existing gender inequalities and compound intersecting forms of discrimination" such as those experienced by women living in poverty or women who belong to racial, ethnic, or religious minorities.[79] Inasmuch as gender inequalities limit women's and girls' access to resources or control over decision-making, they are "more likely to be exposed to disaster-related risks and losses" and less able to adapt to climate change conditions.[80] Women and girls also suffer higher rates of morbidity and mortality in the aftermath of disasters, and face a heightened risk of gender-based violence both during and following disasters.[81]

In a 2018 report, the then-UN Special Rapporteur on extreme poverty and human rights stated that climate change will "exacerbate existing poverty and inequality," and will "have the most severe impact in poor countries and regions, and the places poor people live and work."[82] According to the

---

[74]   Comm. on the Rts. of the Child, *General Comment No. 15 (2013) on the right of the child to the right of the highest attainable standard of health (art. 24)*, ¶ 50, U.N. Doc. CRC/C/GC/15, Apr. 17, 2013.

[75]   Mapping Report, *supra* note 9, ¶ 69.

[76]   *Id.* ¶ 15.

[77]   The Committee monitors states' implementation of their obligations under CEDAW.

[78]   Comm. on the Elimination of Discrimination against Women, *General Recommendation No. 37 (2018) on the Gender-Related Dimensions of Disaster Risk Reduction in the Context of Climate Change*, ¶ 2, U.N. Doc. CEDAW/C/GC/37 (Mar. 13, 2018) [hereinafter *General Recommendation No. 37*].

[79]   *Id.*

[80]   *Id.* ¶ 3.

[81]   *Id.* ¶¶ 4–5.

[82]   Philip Alston (Special Rapporteur on Extreme Poverty and Human Rights), *Climate Change and Poverty*, ¶ 11, U.N. Doc. A/HRC/41/39 (June 25, 2019) [hereinafter *Climate Change and Poverty*].

report, "Since 2000, people in poor countries have died from disasters at rates seven times higher than in wealthy countries."[83] Marginalized communities in both rich and poor countries are also disproportionately affected by extreme weather events—both because they live in areas that are more susceptible to climate change and are less able to protect themselves or recover from the crisis,[84] and because they are often overlooked in the context of relief operations. This lack of "disaster justice" was on clear display in the aftermath of Hurricanes Katrina and Maria in the United States.[85] The Special Rapporteur on extreme poverty and human rights also cautioned that an "over-reliance on the private sector" to deliver basic services and social protection "could lead to a climate apartheid scenario in which the wealthy pay to escape overheating, hunger, and conflict, while the rest of the world is left to suffer."[86]

The mass movement already being caused by climate change also has profound and disproportionate impacts on human rights. An estimated 20 million people per year are already displaced as a result of climate change.[87] That's an average of one person every 2 seconds who is forced to migrate, leaving behind their homes, their sources of livelihood, and likely their cultural heritage. The majority of climate migrants live in poor countries and most are internally displaced.[88]

In the end, those who stand to suffer the most have, ironically, contributed the least to the climate crisis. As noted by Amy Sinden,

> In one sense, all of humanity is in this together. But in another sense this crisis divides us both in terms of culpability and vulnerability. The haves of the world are

---

[83]   *Id.* ¶ 10.

[84]   *Id.* ¶ 12.

[85]   Reilly Morse, *Environmental Justice Through the Eye of Hurricane Katrina*, 2008 JOINT CTR. FOR POL. AND ECON. STUD. HEALTH POL'Y INST., 4 (examining "a common history of discrimination in settlement and other living conditions" that disproportionately increased the vulnerability of African-Americans and low-income communities to disaster, as well as the barriers they faced in precaution and recovery). *See also*, Gustavo A. García-López, *The Multiple Layers of Environmental Injustice in Contexts of (Un)natural Disasters: The Case of Puerto Rico Post-Hurricane Maria*, 11 ENVTL. JUST. 101, 103 (2018) (explaining that "Puerto Rico has many '[environmental justice] communities,' which have been subjected to environmental racism," and that the "effects of [Hurricane] Maria were shaped by these pre-existing relations of injustice….").

[86]   *Climate Change and Poverty*, *supra* note 82, ¶ 50.

[87]   OXFAM, *Forced from home: climate-fuelled displacement*, OXFAM INT'L, 1 (Dec. 2, 2019), https://www.oxfam.org/en/research/forced-home-climate-fuelled -displacement.

[88]   *Id.* at 1–2. Asia is home to 80% of those displaced over the last decade. Small island developing states are particularly at risk. *Id.* at 1.

responsible for the vast majority of the greenhouse gases that have already accumulated, and yet it is the have-nots who are likely to bear the brunt of its effects.[89]

The depth of these divisions was acutely brought to light by a 2020 report by Oxfam International entitled *Confronting Carbon Inequality*, which found that between 1990 and 2015—a critical 25-year period when "humanity doubled the amount of carbon dioxide in the atmosphere"[90]—the richest 1% of the world's population was "responsible for more than twice as much carbon pollution as the 3.1 billion people who made up the poorest half of humanity."[91] Moreover, the "richest 10 percent accounted for over half (52 percent) of the emissions added to the atmosphere between 1990 and 2015."[92] As such, and as Carmen Gonzalez explains, "climate change is a paradigmatic example of *distributive injustice.*" Climate change is also "inextricably linked to broader *social injustice,* including an economic order that systematically exacerbates poverty and inequality while exceeding the limits of the planet's finite ecosystems."[93]

III   STATES' HUMAN RIGHTS OBLIGATIONS TO ADDRESS CLIMATE CHANGE

As the effects of the climate crisis become more immediate and clear, so too does the recognition that states have concrete human rights obligations to address environmental harms, and that in turn, human rights must be at the center of states' responses to climate change.[94] Only recently, however, has international law recognized that the human rights and environmental fields are intertwined. Although many of the rights enshrined in the 1948 Universal

---

[89]   Amy Sinden, *Climate Change and Human Rights*, 27 LAND RES. & ENV'T L. 255, 256 (2007).

[90]   Press Release, Anna Ratcliff, OXFAM Int'l, Carbon Emissions of Richest 1 Percent More Than Double the Emissions of the Poorest Half of Humanity (Sept. 21. 2020),   https://www.oxfam.org/en/press-releases/carbon-emissions-richest-1-percent -more-double-emissions-poorest-half-humanity.

[91]   *Id.*

[92]   *Id.* For the full report, *see* Media Briefing, Tim Gore et al., OXFAM Int'l, Confronting Carbon Inequality: Putting Climate Justice at the Heart of the COVID-19 Recovery (Sept. 21, 2020), https://oxfamilibrary.openrepository.com/bitstream/handle/ 10546/621052/mb-confronting-carbon-inequality-210920-en.pdf.

[93]   Carmen G. Gonzalez, *Racial Capitalism, Climate Justice, and Climate Displacement,* 11 OÑATI SOCIO-LEGAL SERIES 108, 113 (2021).

[94]   *See, for example,* G.A. Res. 41/21, (July 23, 2019) ("Reaffirming the United Nations Framework Convention on Climate Change and the objectives and principles thereof, and emphasizing that parties should, in all climate change-related actions, fully respect human rights").

Declaration on Human Rights cannot be meaningfully achieved without a healthy natural environment, the word "environment" is nowhere to be found in the Declaration; nor are environmental concerns reflected in the ICCPR or the ICESCR. In the mid-twentieth century, the environment was simply not at the forefront of the international community's concerns.

It was not until 1972 that the UN held its first major conference on international environmental issues. The UN Conference on the Human Environment (also known as the Stockholm Conference) "marked a turning point in the development of international environmental politics."[95] The Declaration and Action Plan adopted at the Stockholm Conference have shaped nearly every international environmental conference and multilateral environmental agreement since. Although Principle 1 of the Stockholm Declaration declares that "Man has the fundamental right to freedom, equality and adequate conditions of life, in an environment of a quality that permits a life of dignity and well-being,"[96] international human rights law and international environmental law have largely developed along separate tracks, and the right to a healthy environment has yet to be formally legally recognized under international law, although there now appears to be significant momentum in that direction, as discussed further below.

With the appointment by the UN Human Rights Council of an Independent Expert on human rights and the environment in 2012 (now called a Special Rapporteur), the two fields began to meaningfully converge. In his work as the inaugural officeholder for the mandate, John Knox significantly advanced the notion that human rights and environmental protection are interdependent in that one's ability to enjoy the rights to health and life, among numerous other rights, depends on living in a healthy natural environment. Conversely, the exercise of human rights—namely, the rights to receive information about and participate in environmental decision-making processes—can help protect the environment.[97] These connections were further cemented with the 2015 adoption of the Paris Agreement to the UN Framework Convention on Climate

---

[95]   *Report of the United Nations Conference on the Human Environment*, U.N. Doc A/CONF.48/14/Rev.1 (July 25, 1995).
[96]   U.N. Conference on the Human Environment, *Stockholm Conference*, at Principle 1, U.N. Doc. A/CONF.48/14/Rev.1 [hereinafter Stockholm Declaration]. The Stockholm Declaration is formally known as the Declaration of the United Nations Conference on the Human Environment. Principle 1 goes on to state: "...and he bears a solemn responsibility to protect and improve the environment for present and future generations. In this respect, policies promoting or perpetuating apartheid, racial segregation, discrimination, colonial and other forms of oppression and foreign domination stand condemned and must be eliminated." *Id.*
[97]   John Knox, *Greening human rights*, Open Democracy (July 14, 2015), https://www.opendemocracy.net/en/openglobalrights-openpage/greening-human-rights/.

Change (UNFCCC). The Agreement's Preamble makes clear that all parties "should, when taking action to address climate change, respect, promote and consider their respective obligations on human rights."[98]

Human rights treaty bodies have also sought to mend this fragmentation of international law. In its General Comment No. 36 regarding states' obligations to ensure the right to life under Article 6 of the ICCPR, for example, the Human Rights Committee states that "The obligations of States parties under international environmental law should [] inform the content of article 6 of the Covenant, and the obligations of States parties to respect and ensure the right to life should also inform their relevant obligations under international environmental law."[99]

For more than a decade now, the Human Rights Council, treaty monitoring bodies, special rapporteurs, and regional human rights bodies have all called increasing attention to states' obligations to take concerted action to address climate change,[100] a process that John Knox has called the "greening" of human rights.[101] Today, it is well understood that states have both *substantive* and *procedural* human rights obligations to address climate change, and that they have specific obligations to protect groups who are most vulnerable to environmental harms.

A   Substantive Obligations

States' substantive obligations to address climate change can be grouped into five distinct but overlapping categories. Namely, states must: protect human rights from climate-related harms; mitigate climate change by regulating GHG emissions; cooperate internationally to protect human rights against climate-related harms; address the transboundary impacts of climate change;

---

[98]   See U.N. Framework Convention on Climate Change, *Report of the Conference of the Parties on its Twenty-First Session*, U.N. Doc. FCCC/CP/2015/10/Add.1, Annex (Jan. 29, 2016) [hereinafter Paris Agreement], for the original text of the Paris Agreement Preamble. The Preamble goes on to note that states should take into account "the rights of indigenous peoples, local communities, migrants, children, persons with disabilities and people in vulnerable situations and the right to development, as well as gender equality, empowerment of women and intergenerational equity." *Id.*

[99]   *General Comment No. 36*, *supra* note 51, ¶ 62.

[100]   For a detailed record of climate-related recommendations by U.N. human rights treaty bodies, see Sébastien Duyck & Lucy McKernan, CTR. FOR INT'L ENV'T L. & THE GLOB. INITIATIVE, STATES' HUMAN RIGHTS OBLIGATIONS IN THE CONTEXT OF CLIMATE CHANGE 2018 and its 2019 update Sébastien Duyck, Jolein Holtz & Lucy McKernan, Ctr. For Int'l Env't L. & The Glob. Initiative, States' Human Rights Obligations in the Context of Climate Change: 2019 Update (2019).

[101]   *See supra* note 97.

and ensure that human rights are safeguarded in all mitigation and adaptation activities.[102] Each of these obligations are described in turn below.

To begin, states must protect human rights from the kinds of climate-related harms described in Part II, including by undertaking adaptation measures. In the wake of the influential IPCC special report on the impacts of global warming of 1.5°C above pre-industrial levels, the Committee on Economic, Social and Cultural Rights (CESCR)—which monitors states' implementation of the ICESCR— released a statement underscoring that states have an obligation to respect, protect, and fulfill human rights for all, in a non-discriminatory manner. The Committee added that the "failure to prevent foreseeable human rights harm caused by climate change, or a failure to mobilize the maximum available resources in an effort to do so, could constitute a breach of this obligation."[103]

Second, states must mitigate climate change by regulating GHG emissions within their jurisdiction. As articulated by the OHCHR, "States must act to limit anthropogenic emissions of greenhouse gases… including through regulatory measures, in order to prevent to the greatest extent possible the current and future negative human rights impacts of climate change."[104] Domestically, the most publicized effort to date to use a state's human rights obligations to press for greater mitigation action was the case of *Urgenda Foundation v. Kingdom of the Netherlands*. In a landmark December 2019 decision, the Dutch Supreme Court held that the Dutch government, in line with its human rights obligations, must immediately reduce its greenhouse gas emissions. Specifically, the Court ruled that in accordance with its obligations to ensure the right to life and to respect for private and family life (under articles 2 and 8 of the ECHR, respectively), the Netherlands owes a "duty of care" to protect its citizens from climate change and as such must reduce its greenhouse gas emission by 25% in 2020.[105]

Reflecting on the global significance of the case, the UN High Commissioner for Human Rights stated that the decision confirms that "governments have

---

[102] UNEP, CLIMATE CHANGE AND HUMAN RIGHTS, *supra* note 61, at p. IX.

[103] *Committee releases statement on climate change and the Covenant*, U.N. HUM. RTS. OFF. OF THE HIGH COMM'R (Oct. 8, 2018), https://www.ohchr.org/en/NewsEvents/Pages/DisplayNews.aspx?NewsID=23691&LangID=E.

[104] U.N. HUM. RTS. OFF. OF THE HIGH COMM'R, HUMAN RIGHTS AND CLIMATE CHANGE: KEY MESSAGES 2; *see also General Recommendation No. 37, supra* note 78, ¶ 14 (stating that "Limiting fossil fuel use and greenhouse gas emissions and the harmful environmental effects of extractive industries such as mining and fracking, as well as the allocation of climate financing, are regarded as crucial steps in mitigating the negative human rights impact of climate change and disasters").

[105] *State of the Netherlands v. Urgenda Foundation*, ECLI:NL:HR:2019:2007, 2.3.1-2.3.2, and 9, Judgment (Sup. Ct. Neth. Dec. 20, 2019) (Neth.).

binding legal obligations, based on international human rights law, to undertake strong reductions in emissions of greenhouse gases."[106] *Urgenda* is not the only case to advance the notion that states have human rights obligations to mitigate climate change.[107] But it has perhaps received the most attention because it was a successful case involving a state from the Global North.

In addition to mitigating and adapting to climate change in order prevent foreseeable human rights harms, states must also cooperate internationally to protect human rights against climate-related harms. Climate change is an issue of global concern, not least because emissions from any given state can affect human rights across borders. The language of multiple human rights instruments calls on states to engage in international cooperation. The ICESCR, for example, calls on states' parties to "take steps, individually and through international assistance and co-operation, especially economic and technical, to the maximum of its available resources, with a view to achieving progressively the full realization of the rights recognized in the present Covenant by all appropriate means."[108] On the basis of such language, the CESCR has concluded that states have the following *extraterritorial* obligations to protect and promote economic, social and cultural rights in other countries. States must:

> Refrain from interfering with the enjoyment of human rights in other countries; [t]ake measures to prevent third parties (e.g. private companies) over which they hold influence from interfering with the enjoyment of human rights in other countries; [t]ake steps through international assistance and cooperation, depending on the availability of resources, to facilitate fulfilment of human rights in other countries, including disaster relief, emergency assistance, and assistance to refugees and displaced persons; [and e]nsure that human rights are given due attention in international agreements and that such agreements do not adversely impact upon human rights.[109]

The OHCHR adds that these obligations extend to addressing the extraterritorial impacts of environmental harms, underscoring that these standards are

---

[106] *Bachelet welcomes top court's landmark decision to protect human rights from climate change*, U.N. Hum. Rts. Off. of High Comm'r (Dec. 20, 2019), ohchr.org/EN/NewsEvents/Pages/DisplayNews.aspx?NewsID=25450&LangID=E.

[107] UNEP & Colum. L. Sch. Sabin Ctr. For Climate Change, *Global Climate Litigation Report: 2020 Status Review*, 14 (2020) (providing an overview of global climate litigation cases as of 2020, and noting that an "[i]ncreasing numbers of cases rely [] on fundamental and human rights enshrined in international law and national constitutions to compel climate action.") [hereinafter *Global Climate Litigation Report*].

[108] ICESCR, *supra* note 5, art. 2(1).

[109] *OHCHR 2009 Report, supra* note 59, ¶ 86, citing various CESCR General Comments.

consistent with the international environmental law principle of "common but differentiated responsibilities," which is premised on the idea that as the main contributors of GHG emissions, developed countries have a particular responsibility to mitigate climate change and to assist developing countries in addressing the adverse impacts of climate change.[110] In accordance with this principle, both the UNFCCC and the Paris Agreement call upon developed country parties to assist developing countries to meet the costs of climate adaptation, among other things,[111] but these commitments are not enforceable. For more on states' obligations under international environmental law, see Chapter 1: *The International Climate Change Treaty Regime.*

Related to the obligation of international cooperation is the obligation of states to address the transboundary impacts of climate change, an obligation that is arguably also grounded in customary international law, namely the obligation of states not to cause transboundary environmental harm.[112] While the CESCR has interpreted the ICESCR as including extraterritorial obligations of this nature when it comes to economic, social, and cultural rights, as noted by UNEP, "many developed countries have disagreed with this interpretation, and thus there is not a clear consensus on the extraterritorial application of the convention."[113]

Finally, as part of their substantive obligations, states must ensure that human rights are safeguarded in all mitigation and adaptation activities. The UN Human Rights Council has stressed that states must pursue a human rights-based approach in all policies and activities aimed at addressing climate change. A rights-based approach entails: ensuring that the main objective of climate change policies and programs is to fulfill human rights; evaluating the claims of rights-holders and the corresponding obligations of duty bearers;

---

[110] United Nations Framework Convention on Climate Change art. 3(1), June 4, 1992, 1771 U.N.T.S. 107 ("in accordance with their common but differentiated responsibilities and respective capabilities... developed country Parties should take the lead in combating climate change and the adverse effects thereof.") [hereinafter UNFCCC]; UNEP, Climate Change and Human Rights, *supra* note 61, at 26.

[111] *See* UNFCCC *supra* note 110, arts. 4(3)–(5), and 4(7); and Paris Agreement *supra* note 98, art. 7.

[112] UNEP, Climate Change and Human Rights, *supra* note 61, at 25 (citing the jurisprudence of the International Court of Justice on the matter). See also Stockholm Declaration, *supra* note 96, at Princ. 21 (providing that states have "the responsibility to ensure that activities within their jurisdiction or control do not cause damage to the environment of other States or of areas beyond the limits of national jurisdiction").

[113] UNEP, Climate Change and Human Rights, *supra* note 61, 25. *See also* The ETO Consortium, *Maastricht Principles on Extraterritorial Obligations of States in the Area of Economic, Social and Cultural Rights*, (2013) (a set of principles issued by 40 international law experts in 2011 with the aim of clarifying extraterritorial obligations of states on the basis of standing international law).

developing strategies to build the capacity of rights-holders to claim their rights and of duty bearers to fulfill their obligations; monitoring and evaluating outcomes and processes using human rights principles and standards; and finally, incorporating the recommendations of international human rights bodies to inform each step of the process.[114]

B   Procedural Obligations

States also have procedural obligations with respect to climate change. Namely, states must ensure that the affected public: is adequately informed about the impacts of climate change; is adequately involved in decision-making around mitigation and adaptation measures; and has access to effective legal remedies when their rights are violated.[115] These obligations—which are largely rooted in civil and political rights—also find their corollaries in international environmental law as described below. As such, procedural rights are a "key point of intersection between environmental and human rights law."[116]

On the international environmental law end, Principle 10 of the Rio Declaration provides that "each individual shall have appropriate access to information concerning the environment that is held by public authorities... and the opportunity to participate in decision-making processes." Principle 10 adds that "States shall facilitate and encourage public awareness and participation by making information widely available. Effective access to judicial and administrative proceedings, including redress and remedy, shall be provided."[117]

The Aarhus Convention codifies and elaborates on these protections at the European regional level[118] while the Escazú Agreement, which only

---

[114] See *The impacts of climate change on the effective enjoyment of human rights*, U.N. Hum. Rts. Off. of Hum. Comm'r, https://www.ohchr.org/EN/Issues/HRAndClimateChange/Pages/AboutClimateChangeHR.aspx (last visited Mar. 30, 2021).

[115] UNEP, CLIMATE CHANGE AND HUMAN RIGHTS, *supra* note 61, IX.

[116] UNEP, FACTSHEET ON HUMAN RIGHTS AND THE ENVIRONMENT 2 (2015) [hereinafter UNEP FACTSHEET].

[117] United Nations Conference on Environment and Development, Rio de Janeiro, Braz., June 3-14, 1992, Rio Declaration on Environment and Development, Principle 10, U.N. Doc. A/CONF.151/26/Rev.1, Annex 1 (Aug. 12, 1992).

[118] Formally known as the United Nations Economic Commission for Europe (UNECE) Convention on Access to Information, Public Participation in Decision-Making and Access to Justice in Environmental Matters, the Aarhus Convention was adopted in 1998 and came into effect in 2001. See Convention on Access to Information, Public Participation in Decision-Making and Access to Justice in Environmental Matters, June 25, 1998, 2161 U.N.T.S. 447, 38 I.L.M. 517.

recently came into force, codifies these protections for the Latin American and Caribbean region.[119] The Escazú Agreement also contains specific protections for environmental defenders—an issue of particular significance for the Latin American region where the majority of killings of environmental defenders have taken place in recent years.[120]

Procedural obligations can also be found in the UNFCCC and its Paris Agreement. Article 4 of the UNFCCC, for example, calls on states to publish national inventories of greenhouse gas emissions, to implement and publish national mitigation and adaptation programs, to exchange relevant scientific and technological information, and to promote climate change-related education, training, and public awareness.[121] Article 4 of the Paris Agreement similarly calls on states to "prepare, communicate and maintain successive nationally determined contributions that it intends to achieve."[122]

On the human rights side, states' obligations to keep the affected public informed about the impacts of climate change can be traced to Article 19 of the ICCPR, which recognizes the right of all persons "to seek, receive and impart information."[123] Article 25 of the ICCPR recognizes the right of every citizen to "take part in the conduct of public affairs" and in the government of their country. Various treaty bodies have also affirmed that states have an obligation to facilitate public participation in environmental decision-making,[124] particularly where decisions affect vulnerable groups. Article 2 of the ICCPR enshrines the right to an "effective remedy" for human rights violations as do many other human rights instruments.[125]

The duty to carry out environmental impact assessments (EIAs) can be linked to both the rights to information and public participation.[126] This mechanism is provided for in several multilateral environmental agreements and has

---

[119]   Formally known as the Regional Agreement on Access to Information, Public Participation and Justice in Environmental Matters in Latin America and the Caribbean, the Escazú Agreement was adopted in 2018 and came into force in 2021. *Regional Agreement on Access to Information, Public Participation and Justice in Environmental Matters in Latin America and the Caribbean, opened for signature* Apr. 9, 2018, U.N. Doc. C.N.195.2018, C.N.196.2018 (Apr. 9, 2018).

[120]   *See supra* Part II.

[121]   UNFCCC, *supra* note 110, art. 4(1)(a)(b)(h)(i); *see also id.* art. 6(a)(ii)–(iii) (calling on states to promote and facilitate "[p]ublic access to information on climate change and its effects" and "[p]ublic participation in addressing climate change and its effects and developing adequate responses").

[122]   Paris Agreement, *supra* note 98, art. 4(2).

[123]   ICCPR, *supra* note 4, art. 19 (2).

[124]   UNEP, CLIMATE CHANGE AND HUMAN RIGHTS, *supra* note 61, at 17.

[125]   *See* ICCPR, *supra* note 4, art. 2.

[126]   UNEP FACTSHEET, *supra* note 116, at 3.

also been connected to human rights by regional and international courts. The ACHPR, for instance, "found that failure to conduct an environmental impact assessment contributed to a violation of the right to property" while the ECtHR found that such a failure "contributed to a violation of the right to respect for privacy and home life."[127] In the *Pulp Mills Case*, the International Court of Justice held that states, as a matter of customary international law, have an obligation to conduct and disclose the results of EIAs where there is a risk that a proposed activity "may have a significant adverse impact in a transboundary context, in particular, on a shared resource."[128]

C Obligations to Vulnerable Groups

Finally, states have obligations to protect vulnerable groups from climate-related harms. As noted in Part I, governments must ensure human rights in a non-discriminatory manner, including on the basis of race, color, sex, language, religion, political or other opinion, national or social origin, property, birth, or other status.[129] In addition, states have specific obligations to uphold the human rights of particular groups, including women, children, and Indigenous peoples.

The UN Declaration on the Rights of Indigenous Peoples and the International Labour Organization Indigenous and Tribal Peoples Convention, 1989 (No. 169)[130] together articulate the specific rights of Indigenous peoples. Among the most critical of these rights, and as noted in Part II, states must obtain the free, prior, and informed consent of affected Indigenous peoples before proceeding with projects that affect Indigenous territories.[131] Indigenous peoples and other traditional communities are also seen as key actors in mounting an effective response to climate change. Under the Paris Agreement, parties acknowledge that adaptation action should be guided by "traditional knowledge, knowledge of indigenous peoples and local knowledge systems."[132] To that end, a Local Communities and Indigenous Peoples Platform has been established under the UNFCCC which seeks to "strengthen the knowledge, technologies, practices,

---

[127] *Id.* at 3.

[128] UNEP, CLIMATE CHANGE AND HUMAN RIGHTS, *supra* note 61, at 16, citing *Pulp Mills on the River Uruguay* (Argentina v. Uruguay), 2010 I.C.J. 14, 83, p. 204 (April 20).

[129] *See supra* note 8.

[130] *See generally* Convention (No. 169) Concerning Indigenous and Tribal Peoples in Independent Countries, June 27, 1989, 1650 U.N.T.S. 383.

[131] *See supra* note 63 and accompanying text.

[132] Paris Agreement, *supra* note 98, art. 7(5)

and efforts of local communities and indigenous peoples related to addressing and responding to climate change."[133]

Under the Convention on the Elimination of All Forms of Discrimination against Women, states' parties must protect the rights of women and girls and prohibit all forms of discrimination against them. In its study on "gender-responsive climate action," the OHCHR underscored that "States have legal obligations to implement gender-responsive climate policies that empower women, protect their rights, and address the gendered impacts of climate change."[134] Moreover, OHCHR adds,

> women's unique knowledge and experience, particularly at the local level, in areas such as agriculture, conservation and the management of natural resources means that the inclusion of women in climate action and decision-making processes is not simply a legal and moral imperative, but is also critical to effective and informed action.[135]

The Convention on the Rights of the Child (CRC) recognizes that the "inherent dignity and [] equal and inalienable rights of all members of the human family is the foundation of freedom, justice and peace in the world," and calls on states' parties to ensure that in "all actions concerning children... the best interests of the child shall be a primary consideration."[136] The UNFCCC calls on states' parties to "protect the climate system for the benefit of present and future generations of humankind, on the basis of equity and in accordance with their common but differentiated responsibilities and respective capabilities."[137] And the Paris Agreement calls on states to take action to address climate change in a manner that promotes "intergenerational equity."[138]

Putting these commitments to the test, in September 2019, 16 youth filed a complaint against the five highest greenhouse gas emitting nations that have ratified the CRC and its Third Optional Protocol, which allows children to submit complaints to the UN Children's Rights Committee regarding specific violations of their rights under the CRC. The complaint alleged that these countries—namely Argentina, Brazil, France, Germany, and Turkey—had

---

[133] *Local Communities and Indigenous Peoples Platform*, U.N. CLIMATE CHANGE, https://unfccc.int/LCIPP#eq-1 (last visited Mar. 31, 2021).

[134] U.N. High Comm'r for Hum. Rts., *Analytical Study on Gender-Responsive Climate Action for the Full and Effective Enjoyment of the Rights of Women*, ¶ 31, U.N. Doc. A/HRC/41/26 (May 1, 2019).

[135] *Id.* ¶ 21

[136] CRC, *supra* note 6, pmbl., art. 3(1).

[137] UNFCCC, *supra* note 110, art. 3(1).

[138] Paris Agreement, *supra* note 98, pmbl.

violated the petitioners' rights under the CRC by failing to take sufficient action to reduce their greenhouse gas emissions in response to climate change.

Petitioners asked the Committee, *inter alia*, to find that "by recklessly perpetuating life-threatening climate change, each respondent is violating the petitioners' rights to life, health, and the prioritization of the child's best interests, as well as the cultural rights of the Petitioners from indigenous communities."[139] In October 2021, the CRC dismissed the Petitioners' claims on the basis of improper jurisdiction and failure to exhaust local remedies. Significantly, however, the Committee noted in its decisions that states are legally responsible for failures to prevent foreseeable human rights harms caused by climate change.[140]

D   A Right to a Healthy Environment

The right to a healthy environment—or some expression thereof—has been incorporated by a significant majority of states in their national constitutions and laws and has also been recognized by regional human rights systems.[141] There is now also great momentum toward recognizing this right under international law. In October 2021, the U.N. Human Rights Council adopted reso-

---

[139] Sacchi, et al. v. Argentina, et al., Committee on the Rights of the Child, ¶ 33 (Sept. 23, 2019), https://childrenvsclimatecrisis.org/wp-content/uploads/2019/09/2019 .09.23-CRC-communication-Sacchi-et-al-v.-Argentina-et-al-Redacted.pdf)

[140] CRC, Communication No. 104/2019, *Sacchi et al. v. Argentina*, U.N. Doc. CRC/C/88/D/104/2019, ¶ 10.13 (Oct. 8, 2021); CRC, Communication No. 105/2019, *Sacchi et al. v. Brazil*, U.N. Doc. CRC/C/88/D/105/2019, ¶ 10.13 (Oct. 8, 2021); CRC, Communication No. 106/2019, *Sacchi et al. v. France*, U.N. Doc. CRC/ C/88/D/106/2019, ¶ 10.13 (Oct. 8, 2021); CRC, Communication No. 107/2019, *Sacchi et al. v. Germany*, U.N. Doc. CRC/C/88/D/107/2019, ¶ 9.13 (Oct. 8, 2021); CRC, Communication No. 108/2019, *Sacchi et al. v. Turkey*, U.N. Doc. CRC/ C/88/D/108/2019, ¶ 9.13 (Oct. 8, 2021).

[141] *See* UNEP, Joint statement of United Nations entities on the right to healthy environment (Mar. 8, 2021), https://www.unep.org/news-and-stories/statements/joint -statement-united-nations-entities-right-healthy-environment. (noting that the right to a healthy environment has been recognized by more than 150 UN member states). Regionally, the right has been recognized in the African Charter on Human and Peoples' Rights (*supra* note 22, art. 24); and the Additional Protocol to the American Convention on Human Rights (Additional Protocol to the American Convention on Human Rights in the Area of Economic, Social and Cultural Rights art. 11, Nov. 17, 1999, O.A.S.T.S. No. 69) among other instruments. See The Environment and Human Rights (State Obligations in Relation to the Environment in the Context of the Protection and Guarantee of the Rights to Life and to Personal Integrity: Interpretation and Scope of Articles 4(1) and 5(1) in Relation to Articles 1(1) and 2 of the American Convention on Human Rights), Advisory Opinion OC-23/17, Inter. Am. Ct. H.R. (Nov. 15, 2017).

lution 48/13 to recognize "the human right to a clean, healthy and sustainable environment" and invited the U.N. General Assembly to further consider the matter.[142] This movement has been supported by both the current and former special rapporteurs on human rights and the environment, and is propelled by civil society groups that argue that the recognition of a standalone right will "play a crucial role for the realization of environmental justice for communities exposed to degraded, hazardous or threatening environments," and will "create the foundation for strengthening the environmental policies and legislation of States, providing wider support and legitimacy and thus improving their environmental performance."[143]

IV   THE OPPORTUNITIES AND CHALLENGES OF USING INTERNATIONAL HUMAN RIGHTS LAW TO ADDRESS CLIMATE CHANGE

A   The Opportunities

Human rights law presents both opportunities and challenges for compelling states and other key actors to address climate change. To begin, using a human rights framework makes clear the human toll of the climate crisis and helps makes a strong moral case for state action.[144] A human rights framework also

---

[142] David R. Boyd and John H. Knox (respectively, current and former Special Rapporteur on the Issue of Human Rights Obligations Relating to the Enjoyment of a Safe, Clean, Healthy and Sustainable Environment), *Human Rights Obligations Relating to the Enjoyment of a Safe, Clean, Healthy and Sustainable Environment*, U.N. Doc. A/73/188 (July 19, 2018) (recommending that the General Assembly recognize the human right to a safe, clean, healthy, and sustainable environment). *See* Human Rights Council, *Resolution adopted by the Human Rights Council on 8 October 2021: The human right to a clean, healthy and sustainable environment*, U.N. Doc. A/HRC/RES/48/13 (Oct. 18, 2021).

[143] David R. Boyd and John H. Knox (respectively, current and former Special Rapporteur on the Issue of Human Rights Obligations Relating to the Enjoyment of a Safe, Clean, Healthy and Sustainable Environment), Human Rights Obligations Relating to the Enjoyment of a Safe, Clean, Healthy and Sustainable Environment, U.N. Doc. A/73/188 (July 19, 2018) (recommending that the General Assembly recognize the human right to a safe, clean, healthy, and sustainable environment). Ctr. for Int'l Env't L., The Time is Now!: Global Call for the UN Human Rights Council to Urgently Recognise the Right to a Safe, Clean, Healthy and Sustainable Environment ¶ 10 (2020).

[144] *See also* Sinden, *supra* note 89, at 257 (noting that "Profound moral issues" such as climate change "demand a profound response from law, and as we enter the twenty-first century, human rights is (at least at a rhetorical level) the law's best response to profound, unthinkable, far-reaching moral transgression").

makes visible the fact that while all will be affected by climate change, not all will suffer equally. Rather, the climate crisis will exacerbate existing inequalities, as the greatest burdens will be borne by marginalized communities and nations that have contributed the least to the problem.

By calling on states to ensure that climate change policies are aimed at upholding human rights in a non-discriminatory and participatory manner that attends to the needs of vulnerable groups, the human rights framework centers the voices and lived experiences of those who are far too often sidelined in policymaking processes. Here, the framework's insistence on participatory governance makes room for the possibility that those on the front lines of climate change—who have the greatest understanding of the threat and who by sheer necessity have learned how to be climate resilient—could have a seat at the policymaking table. The exercise of participatory rights also strengthens environmental protection by bolstering civil society participation and its "watchdog role."[145]

The human rights framework carries other benefits as well. As noted by Sumudu Atapattu and Andrea Schapper in the context of discussing the pros and cons of using the human rights framework to seek relief for environmental degradation more generally, using the language, framing, and mechanisms of human rights law can also help galvanize political action and propel social movements. It can also provide guideposts to help policymakers develop rights-respecting climate policies.

The "greening" of human rights has also helped to weave together two distinct branches of public international law in a manner that arguably strengthens protections and clarifies obligations on both ends. From a legal standpoint, and perhaps most significantly, connecting states' climate change-related obligations to their human rights obligations also enables access to legal claims and mechanisms that are not available under international environmental law where citizens typically do not have standing to bring claims against states for their failure to meet their environmental obligations.[146] As referenced above, individuals and civil society groups are increasingly turning to human rights treaty monitoring bodies and regional human rights courts to press their claims. Even where the claim does not succeed, they can help build a moral case for the culpability of high-emitting countries, as was the case with the Inuit petition to the IACHR, which squarely took aim at the culpability of the United States while "giving climate change a human face."[147]

---

[145] Sumudu Atapattu & Andrea Schapper, *Pros and Cons of a Human Rights-Based Approach to Environmental Protection, in* HUMAN RIGHTS AND THE ENVIRONMENT: KEY ISSUES 65 (1st ed. 2019) [hereinafter *Pros and Cons of a Rights-Based Approach*].

[146] *Id.* at 74.

[147] *Id.* at 66.

Human rights cases are also breaking new ground. In January 2020, for example, the UN Human Rights Committee considered the case of Ioane Teitiota, whose application for refugee status in New Zealand was previously rejected. Teitiota had claimed that he should not be sent back to his home country of the Republic of Kiribati as rising sea levels were a threat to his life.[148] Although the Committee rejected Teitiota's claim on the reasoning that the danger he faced was not "imminent,"[149] it did open the door to refugee claims on the grounds of climate change as it underscored that it was unlawful for states to return individuals to countries where their lives may be at risk as a result of climate change.[150]

The world over, youth activists are also pushing the boundaries of international human rights law, hoping to seek relief and compel action in the face of an uncertain future. In the case that went before the Children's Rights Committee, for example, petitioners argued that the CRC must be interpreted in a manner that takes into account respondent states' obligations under international environmental law. Namely, they must "prevent foreseeable domestic and extraterritorial human rights violations resulting from climate change... cooperate internationally in the face of the global climate emergency... apply the precautionary principle to prevent deadly consequences even in the face of uncertainty, and ... ensure intergenerational justice for children and posterity."[151]

At the domestic level, and as noted above, while the *Urgenda* case was perhaps the most publicized effort to date to use a state's human rights obligations to press for greater mitigation action, climate cases in other parts of the world are also successfully advancing the notion that a state's failure to take steps to mitigate climate change violates the rights of its citizens.[152] For the

---

[148] U.N. Hum. Rts. Comm., *Views Adopted by the Committee under Article 5 (4) of the Optional Protocol, Concerning Communication: No. 2728/2016*, ¶¶ 2.1 and 3, UN Doc. CCPR/C/127/D/2728/2016 (Sept. 23, 2020).

[149] *Id.* ¶ 8.4. The Committee reasoned that because the likely timeframe for sea-level rise to render Kiribati uninhabitable was 10–15 years, the government had ample time to take measures to protect or relocate its population. *Id.* ¶ 9.12.

[150] *See id.* ¶ 9.11 (stating the Committee's view that "without robust national and international efforts, the effects of climate change in receiving States may expose individuals to a violation of their rights under articles 6 or 7 of the Covenant, thereby triggering the non-refoulement obligations of sending States. Furthermore, given that the risk of an entire country becoming submerged under water is such an extreme risk, the conditions of life in such a country may become incompatible with the right to life with dignity before the risk is realized").

[151] Sacchi, et al. v. Argentina, et al., *supra* note 139, ¶ 175.

[152] For more on these cases, *see generally Global Climate Litigation Report*, *supra* note 107.

time being at least, these examples suggest that using a state's human rights obligations to enforce its climate-related obligations may prove a useful tool in the hands of climate activists.

B   The Challenge

Using a human rights framework is also rife with challenge. To begin, environmental and human rights policies are not always mutually reinforcing. As discussed in Part II, states' mitigation and adaptation strategies can themselves harm human rights. Similarly, states (and "developing" states in particular) can and do argue that they must be free to pursue development-related activities that may harm the climate in the long term, but are needed to address poverty and ensure the social and economic rights of their populations in the short term. Countries from the Global South have in fact long asserted their "right to development" as an issue of greater material need and moral urgency than environmental concerns. Moreover, despite growing consensus that states have human rights obligations to mitigate climate change, it is still unclear what this means in practice. If these obligations are merely tethered to states' meager nationally determined contributions under the Paris Agreement—as was the case in *Urgenda*—then there is little hope that human rights cases will deliver transformative change.

As human rights lawyers and advocates have long known, the ability to bring human rights claims is also far from a panacea. The few claims that manage to reach regional human rights bodies and international human rights mechanisms have had mixed success. Although they have undoubtedly advanced the notion that human rights and climate change are intertwined, the ability to enforce these judgments and recommendations remains elusive. At the end of the day, while the pronouncements of the Human Rights Committee or the Children's Rights Committee carry great moral authority and are technically binding on states that are parties to the underlying treaties, these Committees lack the power to enforce their decisions, leaving it to the states and civil society actors therein to ensure that the recommendations are implemented. In short, human rights law does not have the power to implement its normative terms, resulting in widespread impunity for violations of rights. Indeed, an essential problem with the human rights framework is that the state is both the target as well as the guarantor of the reforms promoted. As such, civil society actors play a critical enforcement role in pushing states to uphold their human rights obligations.

Problems enforcing rights guarantees on the domestic plane are further compounded by global power dynamics. States' human rights obligations often come into conflict with their investment, trade, or debt-servicing obligations. These conflicts are often resolved in a manner that favors the interests of powerful economic actors. In the end, even as the human rights and environmental branches of public international law are becoming less fragmented, the divisions between public and private international law remain, with the latter wielding far greater power over states' actions.

Beyond enforcement-related concerns, another key challenge of using the human rights framework to address climate change is the fact that human rights law leaves key actors and impacts outside its purview. As explained in Part I, the foundational paradigm of international human rights law is the accountability of sovereign states for ensuring the rights of individuals under their jurisdiction. Implicit in this state-centric approach is the rationale that human rights are the byproduct of relationships between governments and the individuals they govern, rather than relationships between global actors and individuals and communities worldwide whose rights are affected by their actions.[153] In the context of climate change, a variety of state and non-state actors are contributing to climate harms, and that too, over long periods of time. But not all actors are given equal consideration under international human rights law. Specifically, human rights law still does not adequately address the obligations of the private sector or the obligations of states to uphold human rights extraterritorially.

As discussed in Part I, states have an obligation to protect individuals against abuses by third parties, including the private sector. States must also provide access to an adequate remedy where rights are violated, including by non-state actors. Neither international human rights law nor international environmental law, however, imposes direct obligations on the private sector to ensure human rights or address climate change. There are, however, some non-binding or "soft law" standards that states are beginning to reflect in their domestic laws. These standards are embodied in the *Guiding Principles on Business and Human Rights* which were unanimously adopted by the UN Human Rights Council in 2011 and which call on business enterprises, "in all contexts," to "comply with all applicable laws and respect internationally recognized human rights, wherever they operate."[154]

---

[153] Smita Narula, *The Right to Food: Holding Global Actors Accountable Under International Law*, 44 COLUM. J. TRANSNAT'L L. 691, 694 (2006).

[154] John Ruggie, *Guiding Principles on Business and Human Rights: Implementing the United Nations "Protect, Respect and Remedy" Framework*, princ. 23 U.N. Doc. A/HRC/17/31 (Mar. 21, 2011) (endorsed by the U.N. Human Rights Council, Res. 17/4, June 16, 2011) (report of the Special Representative of the Secretary-General on the issue of human rights and transnational corporations and other business enterprises), *available at* www2.ohchr.org/english/bodies/hrcouncil/docs/17session/A.HRC.17.31 _en.pdf (last visited March 21, 2021). This corporate responsibility to respect human rights applies "to all business enterprises, both transnational and others, regardless of their size, sector, location, ownership and structure." *Id.* at princ. 14. Business's respon-

To meet their human rights responsibilities, businesses are called upon to implement policies and processes to safeguard human rights in all aspects of their operations.[155] In the context of climate change, and according to OHCHR, this could mean that they have a policy in place that clearly states their commitment to addressing climate change and the steps that they will take in that regard, including by reducing the emission of GHGs and deforestation that result from their operations or that are directly linked to their products or services. Carrying out environmental and human rights impact assessments should be an integral part of this process and businesses should have in place "processes to enable the remediation of any adverse human rights impacts that they cause or to which they contribute, including through their direct or indirect emissions of greenhouse gases and toxic waste."[156] This corporate responsibility to respect human rights is, however, "distinct from issues of legal liability and enforcement"[157] and translating these responsibilities into concrete efforts by fossil fuel companies to both mitigate and take responsibility for climate change can be fraught.

Nevertheless, human rights bodies are pushing forward to try and assign direct human rights responsibilities.[158] In 2015, for example, the Philippines' Commission on Human Rights (CHR) launched an inquiry into the human rights impacts of climate change on Filipinos and the contribution of so-called "carbon majors" to climate change. The four-year inquiry, which was initiated in response to a petition filed by victims of Typhoon Haiyan and by Greenpeace Southeast Asia, was "the first of its kind to undertake a serious examination of how the world's largest producers of fossil fuels have contributed to climate-related human rights violations."[159] In December 2019, at

---

sibility to respect human rights covers the full range of rights included in the UDHR, the ICCPR, the ICESCR, as well as the eight International Labour Organization core conventions. *Id.* at princ 12.

[155]　*Id.* at Part II.

[156]　OHCHR FAQ, *supra* note 35, at 37.

[157]　*Id.*

[158]　Since 2014, the UN Human Rights Council has taken steps to elaborate an international legally binding instrument to regulate the activities of transnational corporations and other business enterprises with respect to human rights. For more on the treaty process, see *Binding Treaty*, Business And Human Rights Resource Centre, https://www.business-humanrights.org/en/big-issues/binding-treaty/ (last visited Apr. 27, 2021).

[159]　Press Release, Groundbreaking Inquiry in Philippines Links Carbon Majors to Human Rights Impacts of Climate Change, Calls for Greater Accountability, Ctr. for Int'l Env't L. (Dec. 9, 2019) [hereinafter Groundbreaking Inquiry in Philippines], https://www.ciel.org/news/groundbreaking-inquiry-in-philippines-links-carbon-majors -to-human-rights-impacts-of-climate-change-calls-for-greater-accountability/.　For more on the Philippines' Commission on Human Rights' inquiry, see *National Inquiry*

COP25, the CHR announced the results of its investigation, concluding that carbon major companies "could be found legally and morally liable for human rights violations arising from climate change."[160]

But the Commission itself did not set out to adjudicate the responsibility of carbon major companies, nor did it claim the jurisdiction to do so.[161] Rather, it concludes that these cases could and should be brought in domestic courts using existing or newly adopted national laws.[162] Indeed, in the absence of binding international laws and mechanisms to enforce business's responsibility to respect human rights, it will be up to states to hold fossil fuel companies accountable through their national laws. In this regard, the Commission did find that "in circumstances involving obstruction, deception, or fraud, the relevant *mens rea* (criminal intent) may exist to hold companies accountable under not only civil but criminal laws."[163]

Related to the issue of holding fossil fuel companies accountable is the issue of holding states accountable for the extra-territorial impacts of their GHG emissions. Although human rights bodies have begun to elaborate on states' extraterritorial obligations to protect human rights, we are still far from turning those pronouncements into concrete levers for action. As noted in Part III, many developed countries disagree that they in fact have any extraterritorial obligations. Even if these accountability hurdles could be overcome, one is still left with complicated questions around causation and the kinds of remedies that might be appropriate to make whole those who have suffered climate-related harms. In the end, and as articulated by Atapattu and Schapper, a "complex constellation of cause and effect over time, with public and private actors involved in different world regions can make identifying duty bearers and assigning (extraterritorial) obligations an ambiguous task."[164]

But perhaps a broader and more profound critique is that the international human rights framework is ill equipped to deliver climate justice in the global sense of the term. Put differently, what good does it do to insist that the government of the small island nation of Vanuatu fulfill its human rights obligations to protect its population from rising sea levels when it has neither caused the problem, nor has sufficient resources to fix the problem? Human rights law's response to this query is to point to the obligation of international cooperation, discussed above, but high carbon-emitting states have so far refused to couch

---

on *Climate Change*, Republic of the Philippines Commission on Human Rights, https://chr.gov.ph/nicc-2/ (last visited Apr. 29, 2021).

[160] *Groundbreaking Inquiry in Philippines*, *supra* note 160.
[161] *Id.*
[162] *Id.*
[163] *Id.*
[164] *Pros and Cons of a Rights-Based Approach*, *supra* note 145, at 76.

their aid-giving actions in legal obligation terms. Ultimately, while the development of international human rights law has enabled individuals to name and claim their rights against states, human rights law itself was not designed to challenge an unjust international economic order.

Finally, the anthropocentric nature of human rights ironically limits the potential of the rights framework to serve human needs. The sustainable fulfillment of human rights is, after all, predicate on our relationship with, and balance within, the ecosystem in which we live. The human rights framework, however, gives primacy to the rights of one species, creating an imbalance as a starting point. To the extent that the human rights framework does value non-human life and the natural world, it does so in instrumental rather than intrinsic terms.[165]

CONCLUSION

Even with its shortcomings, international human rights law and related mechanisms have provided fertile ground for civil society members to press their climate change-related claims and advance their efforts to hold states accountable. Human rights law itself has also evolved to respond to emerging human rights threats and changing climate conditions. As such, the important work of articulating states' international cooperation and extraterritorial human rights obligations, including their responsibilities to regulate fossil fuel companies that are causing global climate harms, will no doubt continue. Efforts to assign direct human rights responsibilities to the private sector must also move forward. Simultaneously, much more needs to be done to develop and enforce far more ambitious climate mitigation targets under international environmental law. In the end, international human rights law is a legal, political, and rhetorical tool that is most effective when powered and enforced by civil society actors and social movements to help advance their claims for social and climate justice. Given the monumental and existential task that confronts us, it is a welcome and ever-sharpening tool in the climate change toolbox.

---

[165] Smita Narula, *The Right to Food: Progress and Pitfalls*, 2 Canadian Food Stud. 41, 44–45 (2015). The Universal Declaration of the Rights of Mother Earth states that "in an interdependent living community it is not possible to recognize the rights of only human beings without causing an imbalance within Mother Earth... to guarantee human rights it is necessary to recognize and defend the rights of Mother Earth and all beings in her...." The Declaration was adopted in 2010 in Bolivia at the World People's Conference on Climate Change and the Rights of Mother Earth. For more on the Declaration, see *Universal Declaration of Rights of Mother Earth*, Glob. All. for the Rts. of Nature, https://www.therightsofnature.org/universal-declaration/ (last visited Apr. 13, 2021).

7. Legal and policy levers to prompt action by private climate change actors

Katrina Fischer Kuh

In the absence of strong governmental limits on greenhouse gas (GHG) emissions despite increasingly apparent climate change harms, those seeking to advance climate change mitigation policy and/or obtain compensation for climate change harms utilize a variety of legal and policy levers to engage directly with private actors whose conduct contributes to climate change. One approach seeks to use litigation—lawsuits against entities that have significantly contributed to climate change (most notably the fossil fuel industry and large energy companies) requesting a court order to reduce the emissions-causing conduct and/or damages for the plaintiff's climate change harms—to directly compel changes to the behavior of or obtain compensation from these private climate change actors. Note, this litigation is distinct from citizen suits to enforce statutory controls; these lawsuits seek to hold private climate change actors accountable for failing to satisfy obligations, primarily grounded in the common law, independent of any legislative mandate. While litigants have brought many cases of this nature in numerous jurisdictions,[1] to date no case has been allowed to proceed to trial, let alone resulted in a judgment against a private climate change actor. That fact, however, understates the import of these lawsuits. The lawsuits call public attention to how some private climate change actors continue to profit from conduct that is causing dire climate change harms—a salient truth even if it ultimately turns out that there is no legal redress for those harmed. Proceeding to trial in a case, even if defendants are ultimately not held to be liable, could allow for additional publicity and important discovery. And if the cases gain traction and become

---

[1]   The cases have primarily been brought in the United States, although one exception is a suit brought by a Peruvian farmer against a German utility company based on its emissions and contribution to climate change, seeking damages for harms from the melting of a glacier. Lliuya v. RWE AG, VG Essen 15.12.2016 (2 O 285/15) (Germany).

onerous to defend and/or defendants perceive a credible threat of a bad judgment—a single damages award could result in staggering liabilities—this might cause them to support federal climate change legislation that would preempt (and thereby extinguish) the claims. And a court order requiring defendants to reduce their emissions and/or pay damages to those harmed by climate change could prove transformative by significantly hastening the transition away from carbon-intensive sources of energy.

The absence of strong federal climate change law and policy also increases the relative importance of private climate change governance. The term private climate change governance means private environmental governance—"incentives to achieve environmental protection ... even in the absence of government action,"[2] including "initiatives taken by the private sector—businesses, civic and advocacy groups, universities, hospitals, religious organizations"[3]— aimed specifically at climate change mitigation or adaptation. Private climate change actors may voluntarily adopt prescriptive standards that they impose on themselves (for example, an internal GHG emission reduction target), become subject to GHG reporting requirements or mitigation requirements through membership in an industry association, or adopt climate-friendly practices to obtain a certification verified by a third-party nongovernment organization (for example, US Green Building Council's Leadership in Energy and Environmental Design (LEED) certification).[4] In some cases, private climate change actors go even further, imposing climate change-friendly requirements on other private sector actors, as, for example, when companies require supply chain contractors to meet reporting or performance standards related to climate change.[5]

---

[2]   Michael P. Vandenbergh, *Private Environmental Governance*, 99 CORNELL L. REV. 129, 163 (2013).

[3]   Michael P. Vandenbergh, Shannon Vreeland, Ted Atwood, *Private Governance Response to Climate Change: The Case of Refrigerants*, NAT. RESOURCES & ENV'T, Spring 2019, at 31, 32. *See also* Michael P. Vandenbergh, *Private Environmental Governance*, 99 CORNELL L. REV. 129, 133 (2013) (describing private environmental governance as "private interactions in social settings and the marketplace" that produce "a new model of legal and extralegal influences on the environmentally significant behavior of corporations and households."). Private environmental governance includes individual and household behaviors, which are treated separately in Chapter 8; the focus in this chapter is the private corporate sector.

[4]   For an overview of forms of private environmental governance, *see* Sarah E. Light & Eric W. Orts, *Parallels in Public and Private Environmental Governance*, 5 MICH. J. ENVTL. & ADMIN. L. 1, 23–29 (2015).

[5]   The Climate Disclosure Project supports an extensive supply chain membership program for interested entities. Climate Disclosure Project, Supply Chain Program, available at https://www.cdp.net/en/supply-chain (last visited December 31, 2020).

Although they are almost certainly not sufficient, standing alone, to provide the backbone of a societal response to climate change, private climate change governance initiatives may prove crucial for limiting damage from the delay in meaningful governmental action on climate change in at least two respects. First, they can achieve near-term emissions reductions in the absence of governmental emission reduction mandates, thereby reducing the emissions burden created by the lag in political action. Second, private climate change governance initiatives can enhance the efficacy and speed of the implementation of climate change laws once adopted by prompting industry and the commercial sector to develop climate change-friendly infrastructure and practices. Notably, navigating the web of private climate change governance initiatives, both in terms of evaluating participation and compliance, constitutes a core task for many lawyers currently advising corporate clients on climate change.

A second broad approach to directly engage private climate change actors works through channels of private climate change governance and seeks to spur and/or support voluntary mitigation or adaptive corporate action. These efforts include everything from education and moral suasion to market pressure (including divestment campaigns and boycotts) and shareholder resolutions. There are numerous levers for prompting voluntary changes in corporate behavior related to climate change, and a thorough review of the myriad voluntary disclosure and certification regimes is beyond the scope of this text.[6] However, as this volume provides an introduction to the law of climate change, it seems fitting to both note the existence and importance of private climate change governance (as a precursor and supplement to government policy) and also to examine an important way in which law can support private climate change governance, specifically by requiring the disclosure of climate change risks and impacts by private climate change actors under federal securities laws.[7]

This chapter first considers efforts to use the law to hold private climate change actors financially accountable for GHG emitting activities that have contributed to climate change and/or to compel them to cease conduct that exacerbates climate change. Can the common law hold large fossil fuel pro-

---

[6]   Some of the more important voluntary, third-party climate change disclosure and/or certification regimes include the Climate Disclosure Project, the Task Force on Climate-Related Financial Disclosures, the International Organization for Standardization, Coalition for Environmentally Responsible Economies (CERES), the Sustainability Accounting Standards Board, and the Climate Disclosure Standards Board.

[7]   For a thorough analysis of divestment strategies and corporate and securities law mechanisms to influence investment in support of climate change mitigation, *see* Hari M. Osofsky et. al., *Energy Re-Investment*, 94 IND. L.J. 595, 609 (2019).

ducers or those who operate power plants that emit large volumes of GHGs accountable to those harmed by climate change—counties paying millions of dollars to prepare for sea-level rise and other climate change impacts, communities forced to relocate when climate change impacts render their communities uninhabitable, fishermen after fisheries crash because of climate change harms to species? The chapter then turns to consider how law can be used to encourage voluntary climate change mitigation by private climate change actors, primarily through disclosures required under federal securities laws.

I CLIMATE CHANGE SUITS AGAINST PRIVATE CLIMATE CHANGE ACTORS

Litigants filed a first generation of lawsuits against private climate change actors—large emitters of GHGs and/or fossil fuel producers—in the United States between 2004 and 2008. These first-generation lawsuits primarily alleged that the defendants' direct emissions or indirect emissions (from products that they manufactured and/or the combustion of fossil fuels that they produced) contributed to climate change, which constituted a public nuisance. The first-generation lawsuits included a suit brought by the State of California against automobile manufacturers seeking damages for climate change harms in California on the grounds that emissions from their vehicle fleets contribute to climate change (*California v. General Motors Corp.*); a suit by individual property owners seeking damages for harm to their property from Hurricane Katrina brought against multiple oil, electric, chemical, and coal companies (*Comer v. Murphy Oil USA, Inc.*) ("*Comer*");[8] a suit brought by the residents of the Native Village and City of Kivalina, Alaska, facing the prospect of relocation as a result of increasing storm damage and flooding against multiple oil, energy, and utility companies (*Native Village of Kivalina v. ExxonMobil Corp.*) ("*Kivalina*"); and a suit brought a group of states, New York City and nonprofit land trusts against five energy companies (*American Electric*

---

[8] The procedural history in *Comer* is painfully convoluted. A district court dismissed the action upon determining that the plaintiffs lacked standing and that the case presented a nonjusticiable political question in 2005. A panel of the Fifth Circuit would have reversed the dismissal and allowed some of the claims to proceed, but the Fifth Circuit panel decision was vacated for rehearing *en banc*, after which the Fifth Circuit lost its quorum, an unusual procedural occurrence which resulted in the initial district court dismissal being allowed to stand. The parties nonetheless attempted to file a new case, which a district court again dismissed in 2012 on a number of grounds, including res judicata and collateral estoppel, and the Fifth Circuit affirmed the dismissal under res judicata and estoppel. References to *Comer* herein are to the second district court dismissal, which, unlike the first district court dismissal, was decided after some important intervening Supreme Court precedent.

Power Co. v. Connecticut) ("*AEP v. Connecticut*"). Courts dismissed every first-generation suit prior to trial. The lawsuits nonetheless impacted climate change policy and future litigation in important respects. Some of the cases, in particular the *Kivalina* case, garnered significant public attention and, by seeking to hold defendants accountable for current and near-term harms from climate change, challenged the then-widespread perception of climate change as a temporally and geographically distant threat. Additionally, the litigants in these first-generation cases tested the application of various legal doctrines in the context of common law climate change suits; the resulting decisions shaped the second-generation climate change cases which are now being adjudicated.

It would be remiss to proceed to consider the legal aspects of these first-generation lawsuits without acknowledging the vision that they embody, the underlying public interest, and the gravity of the harms they seek to prevent and/or redress. It is hard now to appreciate how breathtaking these claims seemed at the time they were filed, both for their novelty and potential impact. In the late 1990s and early 2000s, scholars and climate policy advocates started to raise the theoretical possibility of suing entities for contributing to climate change;[9] by 2004, these theories were being written into complaints. The dimming of the prospects for a timely and efficacious national response to mitigate and adapt to climate change heightens the relative importance and potential value of common law litigation as a means to fill the gap in government action and spur climate change mitigation. And the increasing frequency and devastation of manifested harms from climate change, often visited most severely upon those least responsible, gives urgency to the claims that those responsible should be held accountable. Notably, the residents of Kivalina remain in peril, but they are hardly alone; we have watched homes burn across California and floods displace people from Texas to Florida. While the role of this catalyst litigation in promoting climate change mitigation and compensating those harmed by climate change remains to be seen, the litigation stands in stark contrast to the lack of public and political will to mitigate climate change.

The first generation of climate change suits against private actors raised and provided answers, or at least some preliminary judicial thinking, on four central legal questions: Do plaintiffs have standing to sue private climate change actors for their contributions to climate change? (Maybe.) Are climate change suits against private climate change actors barred under the political question doctrine? (Probably not.) Has the Clean Air Act (CAA) displaced suits against private climate change actors? (Yes, at least those sounding in

---

[9]   *E.g.*, David A. Grossman, *Warming Up to a Not-So-Radical Idea: Tort-Based Climate Change Litigation*, 28 COLUM. J. ENTL. L. (2003).

federal common law and brought against some types of defendants.) Can plaintiffs state a cause of action under public nuisance doctrine? (Probably.)

The first three questions—relating to standing, the political question doctrine, and displacement or preemption—implicate distinct doctrines addressed to the threshold issue of justiciability, i.e., whether claims of this type can be adjudicated in the judicial forum. Only the last question, evaluating whether entities' direct or indirect emissions can be deemed to contribute to a public nuisance (climate change), relates to the merits of the claims and, notably, only one court reached this issue, and its decision was overturned on other grounds. In *AEP v. Connecticut*, the Second Circuit Court of Appeals held that climate change constitutes a public nuisance (an unreasonable interference with a right common to the general public) to which the defendants' emissions contribute, thereby stating a claim for public nuisance. The Second Circuit rejected efforts by the defendant to argue that public nuisance only encompasses nuisances of a simple type, where the harms are localized, the nuisance is toxic, and the connection between the nuisancing conduct and the harm is apparent and direct. As discussed below, the Second Circuit's decision was overruled on other grounds. However, the Second Circuit concluded that the plaintiffs "stated a claim under the federal common law of nuisance," giving future plaintiffs in climate change litigation reason to be encouraged that there is a possibility of substantive relief should they surmount threshold jurisdictional obstacles.[10]

Since dispositions in the first generation of suits against private climate change actors did not definitively resolve whether such claims are, in fact, justiciable, it remains in doubt whether courts can hear such suits. The decisions in the first-generation climate nuisance suits did, however, suggest strategies for styling the second generation of climate nuisance suits to increase the prospects that courts will hear them.

A   Standing

For a plaintiff to have standing to allow a court to hear the plaintiff's claim, the defendant must have caused a concrete and particularized injury to the plaintiff that is actual or imminent and that injury must be redressable by the relief requested by the plaintiff (in short, injury, causation, and redressability). One recurring issue in early climate litigation was whether harms caused by climate change, which may depend upon scientific modeling and manifest well into the future, can be considered actual or imminent for purposes of standing. No court in the first-generation climate change suits held that the

---

[10]   Connecticut v. Am. Elec. Power Co., 582 F.3d 309, 392 (2d Cir. 2009), *rev'd*, 564 U.S. 410, 131 S. Ct. 2527, 180 L. Ed. 2d 435 (2011).

plaintiffs did not have standing because the alleged climate harms were too remote, but it was not unusual for litigants and judges to spend significant time and effort addressing the issue, leaving the impression that it presented a somewhat close question. Although courts often held that future harms from climate change satisfied the relevant standing requirement, they were also careful to identify *current* climate change harm experienced by the plaintiffs. For example, the Second Circuit in *AEP v. Connecticut* emphasized that the California snowpack was already reduced in size in its standing analysis. This caution is understandable as the dissenting justices in *Massachusetts v. EPA*, discussed further in Chapter 5, would have held that the plaintiffs in that case did not have standing in part because "accepting a century-long time horizon and a series of compounded estimates renders requirements of imminence and immediacy utterly toothless."[11] Going forward, however, climate change plaintiffs will likely have little difficulty establishing this aspect of standing in many cases. Climate change harms—from wildfires to flooding—now manifest with disturbing ferocity and frequency; moreover, the science of attribution has advanced significantly, allowing more definitive conclusions about the connection between these events and climate change.[12] Thus, the uncertainty about and extended analysis of whether climate change harms can be deemed imminent seem destined to become a relic of the first-generation climate suits, testament to a time when we had the luxury to think of climate change as a far-away likely outcome, as opposed to a constant condition.

Climate change nuisance plaintiffs must also, however, show that the defendant's complained-of conduct caused their injury and, relatedly, that the relief that they seek will in fact remedy (or fix or solve) their injury. The numerosity of the contributors to climate change (both over time and currently) as well as the attenuation between the challenged acts (emission of GHGs or production of fossil fuels) and the harm to plaintiff (through the manifested impacts of climate change) makes it difficult for climate change nuisance plaintiffs to establish causation. Lower courts reach different conclusions and there are reasons to think that at least some justices of the Supreme Court are skeptical of whether climate change plaintiffs have standing in this context. With respect to causation, the question is whether the plaintiff's alleged harms from climate change are fairly traceable to the GHG-producing conduct of the defendants. The district courts in *Kivalina* and *Comer* held not. The district court judge in *Kivalina* reasoned that "it is not plausible to state which

---

[11]   Massachusetts v. E.P.A., 549 U.S. 497, 542, 127 S. Ct. 1438, 1468, 167 L. Ed. 2d 248 (2007).

[12]   Michael Burger et. al., *The Law and Science of Climate Change Attribution*, 45 COLUM. J. ENVTL. L. 57, 61 (2020).

emissions—emitted by whom and at what time in the last several centuries and at what place in the world—'caused' Plaintiffs' alleged global warming related injuries."[13] The Second Circuit in *AEP v. Connecticut* was willing to connect the dots between the defendants' emissions and the plaintiffs' climate change harms, reasoning that

> [f]or purposes of Article III standing [Plaintiffs] are not required to pinpoint which specific harms of the many injuries they assert are caused by particular Defendants, nor are they required to show that Defendants' emissions alone cause their injuries. It is sufficient that they allege that Defendants' emissions contribute to their injuries.[14]

As of now, it is not clear how the Supreme Court would rule on the question of causation. The Supreme Court signaled significant division on the question of standing, including the precedential value and interpretation of its standing analysis in *Massachusetts v. EPA*, when it upheld the Second Circuit's standing determination in *AEP v. Connecticut* by an equally divided Court, observing that "[f]our members of the Court, adhering to a dissenting opinion in *Massachusetts* ... or regarding that decision as distinguishable, would hold that none of the plaintiffs have Article III standing."[15] Notably, in *Massachusetts v. EPA*, the Supreme Court held that Massachusetts had standing to petition the EPA for stricter emission controls under the CAA. Massachusetts alleged that the EPA's failure to exercise its authority to reduce emissions from vehicles would worsen its harms from climate change. In holding that Massachusetts had standing, the Court emphasized that because Massachusetts invoked a procedural right afforded to it by statute as a sovereign state advancing a quasi-sovereign interest, it would afford "special solicitude" to Massachusetts in its standing analysis.[16] Four dissenting justices rejected the concept of special solicitude and expressed the view that, without it, the plaintiffs could establish neither causation nor redressability because the connection between any emissions reduced by the EPA's action and Massachusetts' alleged injury (sea-level rise) was "far too speculative" such

---

13   Native Vill. of Kivalina v. ExxonMobil Corp., 663 F. Supp. 2d 863, 881 (N.D. Cal. 2009), aff'd, 696 F.3d 849 (9th Cir. 2012).

14   Connecticut v. Am. Elec. Power Co., 582 F.3d 309, 347 (2d Cir. 2009), rev'd, 564 U.S. 410, 131 S. Ct. 2527, 180 L. Ed. 2d 435 (2011).

15   Am. Elec. Power Co. v. Connecticut, 564 U.S. 410, 420, 131 S. Ct. 2527, 2535, 180 L. Ed. 2d 435 (2011).

16   Massachusetts v. E.P.A., 549 U.S. 497, 520, 127 S. Ct. 1438, 1454, 167 L. Ed. 2d 248 (2007).

that it would be "pure conjecture to suppose that EPA regulation of new automobile emissions will likely prevent the loss of Massachusetts coastal land."[17]

It is unclear how the standing precedent in *Massachusetts v. EPA* applies in the context of common law suits where the rationales for extending special solicitude are not present because the plaintiffs do not seek to enforce procedural rights under a statute and/or may not possess the same quasi-sovereign interest. The district courts in *Kivalina* and *Comer* and the Second Circuit in *AEP v. Connecticut* considered and applied *Massachusetts v. EPA* in their standing analyses but reached contrary conclusions, evidencing the malleability of that precedent and underscoring uncertainty about whether plaintiffs can establish standing. Of note, the district court in *Comer*, with the benefit of the Supreme Court's decision in *Massachusetts v. EPA* as well as the Court's subsequent division on standing in *AEP v. Connecticut*, concluded that the plaintiffs in the *Comer* case did not have standing. The district court in *Comer* reasoned that the fact that the plaintiffs had no claim to special solicitude in the standing analysis was dispositive:

> Although it is true that the Supreme Court determined that Massachusetts had standing based on the allegation that the EPA's failure to regulate merely contributed to Massachusetts' alleged injuries, this does not mean that the private citizen plaintiffs in the present case can demonstrate the causal connection standard by showing a mere contribution to similar injuries. If contribution were enough, presumably there would have been no need for the Supreme Court to grant Massachusetts special solicitude in its standing analysis.[18]

Many of the second-generation climate change suits were filed in state court and, if adjudicated in state court, will be governed by state standing law, which is sometimes more permissive than federal standing requirements. Nonetheless, the first-generation climate change nuisance suits signal a flashing yellow sign—proceed with caution!—with respect to whether individuals or entities harmed by climate change have standing to sue private climate change actors who have contributed significantly to the problem.

B   Political Question Doctrine

While no federal appellate court dismissed a first-generation climate change nuisance suit under the political question doctrine, *every* district court presented with a first-generation climate change nuisance suit dismissed it as

---

[17]   Massachusetts v. E.P.A., 549 U.S. 497, 545–46, 127 S. Ct. 1438, 1470, 167 L. Ed. 2d 248 (2007).

[18]   Comer v. Murphy Oil USA, Inc., 839 F. Supp. 2d 849, 861 (S.D. Miss. 2012), aff'd, 718 F.3d 460 (5th Cir. 2013).

nonjusticiable at least in part on the grounds that federal courts do not have jurisdiction over political (as opposed to legal) questions. In short, there are simply some questions that courts are not equipped or authorized to resolve—such questions may be resolved only by the political branches of government. In some regards, the climate change nuisance suits constitute commonplace tort actions (particularly where they seek damages as opposed to injunctive relief). However, both the indeterminacy and breadth of public nuisance doctrine and the recognition that tort liability could indirectly but significantly shape the contours of energy production contribute to judicial concern about whether the climate change nuisance suits constitute political questions. The Supreme Court has set forth factors for identifying when a case presents a nonjusticiable political question, including if there is "a lack of judicially discoverable and manageable standards for resolving" the case or if it would be impossible for the court to decide the case "without an initial policy determination of a kind clearly for nonjudicial discretion."[19] In the context of climate change litigation, courts have sometimes concluded, relying primarily on those factors, that the litigation presents a nonjusticiable political question. The district court in *Kivalina v. ExxonMobil*, for example, reasoned that resolving the plaintiffs' claim would require it to evaluate whether the defendants' emissions were unreasonable (so as to constitute a public nuisance) by weighing the utility of the challenged conduct (the societal value of low-cost energy) against the gravity of the resulting harm (climate change), an inquiry with respect to which there are no standards and that would require it to make a policy decision. Although the plaintiffs sought damages, as opposed to injunctive relief, and as such any court decision would not dictate emission levels or climate change policy, the district court reasoned that the case nonetheless required it to decide who should bear the cost of climate change, in its view an essentially political judgment. Districts courts in *California v. General Motors Corp.* and *AEP v. Connecticut* similarly dismissed climate change nuisance suits on the grounds that they presented a nonjusticiable political question. Referencing with approval the district court's dismissal in *AEP v. Connecticut*, the district court judge in *California v. General Motors Corp.* concluded that "the adjudication of Plaintiff's claim would require the Court to balance the competing interests of reducing global warming emissions and the interests of advancing and preserving economic and industrial development," which "is the type of initial policy determination to be made by the political branches, and not this Court."[20]

---

[19] Baker v. Carr, 369 U.S. 186, 210, 82 S.Ct. 691, 7 L.Ed.2d 663 (1962).
[20] People of State of California v. Gen. Motors Corp., No. C06-05755 MJJ, 2007 WL 2726871, at *8 (N.D. Cal. Sept. 17, 2007).

Despite these lower court decisions, it seems unlikely that the political question doctrine will result in the dismissal of the second-generation climate change nuisance suits. This is in part because the second-generation cases may proceed in state court, where the federal political question doctrine is inapplicable (and any parallel state doctrines less robust). Even if the cases are successfully removed to federal court, judges concerned about judicial overreach may be more comfortable grounding dismissals in preemption doctrine, as discussed in greater detail below. Additionally, litigants in the second-generation climate change nuisance suits argue that fact finders need not balance social utility and harm—the aspect of the claim identified by courts in the first-generation suits as requiring an impermissible political judgment— to evaluate whether the defendant's conduct constitutes a public nuisance, suggesting instead that unreasonable harm to the plaintiff is sufficient. Notably, in *AEP v. Connecticut*, the Second Circuit focused on evaluating the existence of a public nuisance under the Restatement (Second) of Torts § 821B, which does not directly require utility/harm balancing.

The passage of time and intervening events also cast the climate change nuisance suits in a more benign light that makes them look more like plain vanilla tort suits than vehicles for dictating federal climate change policy that would disrupt the social order. The shift from a top–down to a bottom–up international climate change regime, discussed in greater detail in Chapter 1, blunts concerns that the adjudication of domestic tort suits will undermine international climate change negotiations. Continued revelations about how fossil fuel defendants in these cases purposefully misled the public about climate change suggest how the conduct of those defendants might be understood to be unreasonable (for purposes of showing a public nuisance) even if cheap energy is a valuable public good. And the diminished value and stature of many of the defendant fossil fuel and energy interests makes the possibility that adverse judgments could contribute to causing some of these interests to go bankrupt seem less cataclysmic.

C   Displacement and Preemption

The political question doctrine deems some inquiries as inherently and necessarily fit for resolution only by legislatures (thereby preventing judicial involvement); displacement and preemption doctrines similarly deem some inquiries as *already decided* by legislatures in the form of a statute (thereby limiting judicial involvement to application of the relevant statute). The clearest precedent resulting from the first-generation climate change nuisance suits is the Supreme Court's holding in *AEP v. Connecticut*: "We hold that the Clean Air Act and the EPA actions it authorizes displace any federal common-law right to seek abatement of carbon-dioxide emissions from fossil-fuel fired powerplants."[21] In *AEP v. Connecticut*, a group of states, New York City, and nonprofit land trusts sued five major electric power companies, the five largest emitters of carbon dioxide in the United States, seeking a judgment that their emissions contribute to climate change, a public nuisance in violation of the federal common law of interstate nuisance, or, in the alternative, of state tort law, and requesting an injunction requiring the defendants to reduce their emissions. As alluded to above, the district court in *AEP v. Connecticut* dismissed the case as presenting a nonjusticiable political question. The Second Circuit disagreed, holding that the case did not present a political question. The Second Circuit would have allowed the case to proceed to trial after also concluding that the plaintiffs had standing, the claims were not displaced by the CAA, and the complaints stated a cause of action sounding in public nuisance. Notably, because it determined that the federal common law of public nuisance applied, the Second Circuit did not address the plaintiffs' claims based on state public nuisance law.

The Supreme Court then granted *certiorari* and held that the plaintiffs' federal common law nuisance claims were displaced by the CAA. Because Congress gave the EPA the statutory authority to regulate the emission of GHGs from the defendant power companies (as the Court confirmed in *Massachusetts v. EPA*), that statute (and not federal common law) governed the GHG emissions. The Supreme Court declined to address the plaintiffs' state law public nuisance claims (which had not been briefed), instead remanding them.[22] It observed that because the federal common law nuisance claims were displaced, "the availability *vel non* of a state lawsuit depends, *inter alia*, on the preemptive effect of the federal Act."[23]

The Supreme Court's decision in *AEP v. Connecticut* marks a dead end for at least one type of climate change lawsuit against private actors—a claim under federal common law asking a court to require emission reductions from entities whose emissions are subject to regulation under the CAA. The

---

[21]   Am. Elec. Power Co. v. Connecticut, 564 U.S. 410, 424, 131 S. Ct. 2527, 2537, 180 L. Ed. 2d 435 (2011).

[22]   On remand, plaintiffs obtained leave to voluntarily withdraw their complaints.

[23]   Am. Elec. Power Co. v. Connecticut, 564 U.S. 410, 429, 131 S. Ct. 2527, 2540, 180 L. Ed. 2d 435 (2011). By the time the Supreme Court decided *AEP v. Connecticut*, an appeal from the district court's dismissal in *Kivalina v. ExxonMobil Corp.* was pending before the Ninth Circuit Court of Appeals. The plaintiffs attempted to distinguish *AEP v. Connecticut* and persuade the Ninth Circuit that their federal common law climate change nuisance claim was not similarly displaced by the CAA, in part because they sought monetary damages as opposed to injunctive relief. The Ninth Circuit, however, interpreted Supreme Court precedent to mean that "the type of remedy asserted is not relevant to the applicability of the doctrine of displacement." Native Vill. of Kivalina v. ExxonMobil Corp., 696 F.3d 849, 857 (9th Cir. 2012).

decision, however, leaves open avenues for bringing forward suits grounded in the common law and premised on the same basic underlying facts (a plaintiff harmed by climate change seeking redress from a private actor defendant which has significantly contributed to climate change) but configured in ways more likely to satisfy jurisdictional threshold requirements and therefore to reach fact finders on the merits.

For example, the Supreme Court expressly reserved the question of whether the CAA preempts state common law claims, suggesting the possibility that state common law causes of action might not face the same statutory road-block. The fact that the CAA displaces federal common law does not mean that it also preempts state common law; the legal requirement for establishing preemption is widely recognized as more demanding than that for establishing displacement. A federal statute displaces federal common law where the statute speaks directly to the claim; a federal statute preempts state common law only where there is clear and manifest congressional purpose to do so. Indeed, lower courts have held myriad state common law claims not to be preempted by environmental statutes, including the CAA in other contexts.[24] Additionally, the Supreme Court reasoned that the CAA displaced claims against the large emitters (domestic powerplants) of GHGs because it empowers the EPA to impose limits on emissions from those entities. However, there are numerous potential climate change defendants whose conduct that contributes to climate change is not clearly subject to regulation under the CAA, for example because it involves the production of fossil fuels as opposed to their combustion or because the emissions do not occur in the United States. As discussed below, litigants are actively exploring these potential avenues for stating viable causes of action in a second generation of climate change suits.

D   Second-generation Climate Change Litigation

The second generation of common law climate change suits against private climate change actors differs from its first-generation predecessors in key respects, reflecting, in part, strategic choices informed by the outcomes in the first-generation cases. (Of note, just as there is some diversity in the first-generation cases, there are even more numerous second-generation cases presenting myriad issues, but some broad shifts in the styling and focus of the first- and second-generation cases are apparent.) Most importantly, the plaintiffs in the second-generation climate suits sue under state law. Should

---

[24]   The district court in *Comer* held, with negligible analysis, that the CAA preempted plaintiffs' state law claims. Comer v. Murphy Oil USA, Inc., 839 F. Supp. 2d 849, 865 (S.D. Miss. 2012), aff'd, 718 F.3d 460 (5th Cir. 2013).

the cases ultimately proceed in state courts (or in federal court, but apply state law), differences in the substantive law of different states will shape the cases and may influence outcomes.

The second-generation suits are also more likely to name as defendants fossil fuel producers (for example, oil companies) as opposed to those who directly emit GHGs and may be subject to regulation under the CAA (for example, the owners of coal-fired plants). These shifts could help to avoid the dismissal of the claims as displaced or preempted by the CAA, which imposes controls on the emission of GHGs. The second-generation suits also tend to emphasize relief in the form of damages. Although some of the complaints reference the possibility of injunctive relief to curtail the nuisance, the focus of the lawsuits is on damages. Seeking damages as opposed to injunctive relief may help plaintiffs sidestep other potential grounds for dismissal. Where a plaintiff seeks damages, it may be harder to characterize the plaintiff's harm as not redressable for purposes of standing. Claims seeking damages as opposed to injunctive relief may also be more easily perceived as run-of-the-mill tort suits that will not unduly interfere with federal climate policy and that do not, therefore, present a nonjusticiable political question.

Additionally, the second-generation lawsuits, while typically still alleging nuisance, often invoke other theories of recovery, including most notably strict product liability. With respect to product liability, plaintiffs allege that the fossil fuels sold by defendants were defective in design because of the harms resulting from their combustion (GHG emissions that contributed to climate change) and the failure of defendants to provide consumers with information about those harms. One of the more interesting legal questions relating to the merits of the second-generation suits is whether a product manufacturer may be held liable in public nuisance when its product, when used by third parties, inevitably causes harm. The second-generation lawsuits also tend to center the defendants' knowledge about how their products contribute to climate change and the defendants' efforts to sow doubt about important aspects of climate change science, including whether climate change is occurring and anthropogenic.

It is also worth noting that, while one second-generation case has been brought by an industry trade group (the Pacific Coast Federation of Fishermen's Associations, which alleges harms to the fishery from climate change), it remains more common, as in the first-generation cases, for suits to be brought by governmental entities. In the second-generation lawsuits, the plaintiffs are often local (municipal, county) governments. This may reflect, in part, the continued uncertainty about whether climate change plaintiffs have standing and a judgment that it is easier for governmental entities to show causation and establish standing across larger aggregated geographic areas and timeframes. For example, Marin County's complaint describes the effects of existing and future projected sea-level rise along its lengthy bay- and ocean-adjacent coasts, alleging near-term harm with sea-level rise of 10 inches to 2,000 parcels of land and 800 buildings and longer-term harm with sea-level rise of 5 feet to over 8,000 parcels and 9,000 buildings. Two intervening developments may help plaintiffs in the second-generation suits tie their harms to the defendants' conduct: the publication of the Carbon Majors Report and supporting analyses, which quantify and trace the historic and cumulative emissions of carbon dioxide and methane to the largest fossil fuel and cement producers;[25] and continued advances in the field of climate change attribution, which evaluates whether climate change causes or contributes to specific phenomena (for example, hurricanes).

As of this writing, most of the second-generation lawsuits have been mired in wrangling over whether state or federal law applies to the claims and whether the claims will be heard in state or federal court. Plaintiffs predominantly filed the cases in state courts and defendants removed them to federal court; litigation to date has often centered on the propriety of those removals. Although those federal appellate courts that have heard appeals from removal decisions have unanimously held the removals improper and remanded the cases to state court, the Supreme Court granted *certiorari* in *Mayor & City Council of Baltimore v. BP p.l.c.* to consider the scope of appellate review of district court remand orders. And the Second Circuit Court of Appeals (in a case filed in federal court but alleging violations of state law) and one district court have held that the plaintiffs' claims are necessarily governed by federal common law and therefore displaced by the CAA (with respect to domestic emissions) and unfit for judicial resolution because of extraterritorial effect and undue interference with foreign policy (with respect to emissions outside of the US).[26]

While it seems likely that at least some of the second-generation climate change suits will proceed in state court, it remains uncertain whether they will ultimately survive other jurisdictional challenges and motions to dismiss to be heard on the merits. The first generation of climate change suits evidence

---

[25]   The Carbon Majors Report and related update and analyses are available from the Climate Accountability Institute, https://climateaccountability.org/index.html (last visited December 21, 2020) ("This project is a first attempt at aggregating historic data by carbon producing entities. The work is unique in converting production of all fossil fuels into the carbon content and resulting emissions of carbon dioxide upon the combustion of marketed fuels, and in tracing emissions to the primary producing entities.").

[26]   City of New York v. Chevron Corp., No. 18-2188, 2021 WL 1216541 (2d Cir. Apr. 1, 2021); City of Oakland v. BP P.L.C., 325 F. Supp. 3d 1017, 1025 (N.D. Cal. 2018), *vacated and remanded sub nom.* City of Oakland v. BP PLC, 960 F.3d 570 (9th Cir. 2020), *opinion amended and superseded on denial of reh'g*, 969 F.3d 895 (9th Cir. 2020).

the judiciary's deep unease about overstepping its bounds in the context of climate change policy. The doctrines courts used to dismiss the first-generation cases—standing, political question, displacement—are rooted in limiting federal judicial power to cases and controversies within the constitutional authority and institutional competence of federal courts. State courts may exhibit less judicial restraint, particularly now that the concept of a climate change lawsuit seems less novel. Additionally, the fact that cases are pending in numerous jurisdictions increases the prospect for at least one case to proceed to trial.

Should that happen—should a state law climate change suit go to trial— what would be the consequences? As suggested above, the fact that a case proceeds past threshold jurisdictional obstacles and dispositive motions would provide a powerful incentive, even prior to trial, for current and future defendants to seek statutory protection from liability (i.e., a federal law, perhaps contained as a provision on a larger climate change law, extinguishing common law suits for climate change damages). Indeed, many are already lobbying for such protection. The numerosity of the climate change cases and the diversity of the jurisdictions in which they have been brought means that a verdict for defendants in one or even most cases will not be determinative, as the defendants could still be held liable in other actions. But a verdict *against* the defendants in even a single case would expose those defendants to substantial damages while encouraging more plaintiffs to file suit. And the number of potential plaintiffs inexorably expands as climate change harms manifest and our ability to attribute harms to climate change grows. A successful lawsuit, and those that follow, would deliver a serious financial blow to defendants. As many defendants face pressure on multiple fronts related to their contribution to climate change, including not least market pressure from renewable energy competitors, it is not inconceivable that the climate change suits could contribute to the bankruptcy of one or more defendant companies.

In terms of whether and how these suits contribute to advancing climate change mitigation, one salient issue may be the timeline. Thus far, the defendants have succeeded in significantly slowing down the litigation through disputes over removal; even once those disputes are resolved, lengthy churn over other threshold jurisdictional issues seems likely. Although the suits certainly put pressure on major private actors contributing to climate change, public policy and market forces may cause those actors to change course before the cases reach final resolution. With respect to compensation for climate change harms, however, the timeline may be relevant only insofar as the financial weal of the defendant companies changes over time (thereby affecting their ability to satisfy an adverse judgment). These suits may be the only realistic opportunity for at least some of those harmed by climate change to obtain compensation from at least some of those most culpable for it.

II MANDATED DISCLOSURE AS A TOOL FOR PROMPTING BEHAVIOR CHANGE BY PRIVATE CLIMATE CHANGE ACTORS

In addition to serving as a cudgel to compel accountability, law may serve as a prod to influence voluntary changes by private climate change actors. Many of the tools for influencing the conduct of private climate change actors rely upon transparency. Disclosure related to climate change may work in a self-reflexive manner, as when a corporation is required to develop and report information about its exposure to climate risk and then, newly aware of that risk, takes measures to reduce it. Disclosure can also catalyze corporate behavior change in response to actual or anticipated public or market reaction to disclosed information. Disclosure of emissions and climate risks can inform decisions by investors motivated either by a desire to make financially sound investments and therefore wishing to avoid companies exposed to climate change risks or investors committed to environmental principles. Many large institutional investors commit to follow environmental, social, and governance (ESG) platforms,[27] and a growing divestment campaign seeks to divert investment away from fossil fuel companies. A chief legal means of requiring transparency is through mandated disclosure under securities laws. As discussed below, the requirements for the disclosure of climate change-related information under these laws are rapidly evolving.

Congress has not amended securities laws to require climate change-specific disclosure. However, as with other statutes of general application, like the CAA as discussed in Chapter 2, the securities laws as written require some disclosures relating to climate change, although the precise contours of the legally required disclosure remains unclear.

In 2010, the Securities and Exchange Commission (SEC) issued an interpretive release identifying four items of Regulation S-K that could require disclosure regarding climate change (Item 101, Description of business; Item 103, Legal proceedings; Item 105, Risk factors;[28] and Item 303, Management's discussion and analysis of financial condition and results of operation (MD&A)).[29] Item 101 requires *inter alia* the disclosure of the material effects of complying with regulations concerning the environment

---

[27]   For an overview of investor ESG platforms, *see* Michael P. Vandenbergh et. al., *The Gap-Filling Role of Private Environmental Governance*, 38 Va. Envtl. L.J. 1, 36 table 2 (2020).

[28]   The guidance speaks to Item 503(c), which was relocated to new Item 105 through the Fast Act Modernization and Simplification of Regulation S-K in 2019.

[29]   Commission Guidance Regarding Disclosure Related to Climate Change, Release No. 33-9106 (Feb. 2, 2010) [75 FR 6290 (Feb. 8, 2010)].

(including, for example, state climate change laws) and the material estimated capital expenditures for environmental control facilities.[30] Item 103 requires the disclosure of material pending legal proceedings, including judicial or administrative proceedings relating to discharges to the environment and environmental protection such as the climate change nuisance suits discussed above.[31] Item 105 requires material factors that make an investment in the registrant or offering speculative or risky,[32] such as climate change risks resulting from unfavorable weather conditions to a company that manages and owns ski resorts.[33] Item 303 requires the disclosure of material events and uncertainties known to management that are salient to understanding a registrant's future operating results or financial condition,[34] including known trends, events, and uncertainties, and might include climate change risks. Ultimately, however, whether and how information about climate change must be disclosed depends upon whether the information is material, i.e., whether it is information that is substantially likely to significantly alter the total mix of information available from the perspective of a reasonable investor. The release doesn't provide detailed guidance on this question and neither subsequent SEC reviews of disclosures relating to climate change nor case law have further brought the question into focus. The SEC's silence results in significant uncertainty about the requirements for climate change disclosure and helps to explain the very uneven reporting in practice by many companies.

Although other countries have adopted detailed financial disclosure requirements for sustainability risks, including climate change,[35] and there is clear investor interest in improved and standardized climate change disclosure, the SEC has declined to specify further (beyond the 2010 guidance) what climate change disclosures Regulation S-K requires, either through guidance or use of its disclosure review process, instead encouraging registrants to orient their disclosures to the general principle of materiality. The absence of clearer instruction from the SEC, uneven disclosure by companies in practice, and growing interest in reliable climate change information, including from major investors, has prompted the development of important third-party financial

---

[30]   17 C.F.R. § 229.101(c)(1)(xii) (2020).
[31]   17 C.F.R. § 229.103 (2020).
[32]   17 C.F.R. § 229.105 (2020).
[33]   Roshaan Wasim, Note, *Corporate (Non)disclosure of Climate Change Information*, 119 COLUM. L. REV. 1311, 1324 (2019) (describing disclosures of risk by Vail Resorts, Inc.).
[34]   17 C.F.R. § 229.303 (2020).
[35]   The European Union Regulation on sustainability (environmental, social and governance (ESG)) disclosures in the financial services sector came into force in December 2019.

climate change reporting regimes. These private, voluntary reporting regimes represent a wholly private governance approach, functioning independent of government-mandated disclosure or prescriptive requirements. The Task Force on Climate-Related Financial Disclosures, for example, has authored detailed recommendations for voluntary climate-related financial disclosures[36] and is the framework that major private equity firm BlackRock expects portfolio companies to use to disclose climate-related risks.[37]

State attorneys general and investors have also brought investigations and lawsuits under state securities and consumer protection laws alleging that companies' climate change-related disclosures (or lack thereof) violate those laws. New York has taken a leading role, bringing investigations against multiple energy companies under the Martin Act, which resulted in settlements pursuant to which the companies agreed to disclose certain climate change risks and information.[38] New York also unsuccessfully sued ExxonMobil under the Martin Act and Executive Law § 63(12), alleging that it had misled investors about climate change-related risks to its business, as evidenced in particular by alleged differences between its public representations and internal analyses relating to the proxy cost for carbon.[39] A securities fraud class action lawsuit filed against ExxonMobil Corp., likewise related to its alleged material misrepresentations concerning the proxy cost it used for carbon, is pending in federal district court in Texas.[40]

Going forward, the Biden administration seems likely to prompt the SEC to offer additional guidance about the requirements for climate change disclosure. This will be informed by the third-party reporting regimes that have developed as well as outcomes in the above-described investigations and cases. And establishing robust disclosure requirements for climate change-related risks

---

[36] Task Force on Climate-Related Financial Disclosures, Publications, *available at* https://www.fsb-tcfd.org/publications/ (last visited December 29, 2020).

[37] Additional guidance on climate change reporting, including industry-specific standards, has been developed by the Sustainability Accounting Standards Board and Climate Disclosure Standards Board.

[38] For an overview of the Martin Act investigation and the settlements that resulted, *see* Hana V. Vizcarra, *Climate-Related Disclosure and Litigation Risk in the Oil & Gas Industry: Will State Attorneys General Investigations Impede the Drive for More Expansive Disclosures?*, 43 Vt. L. Rev. 733, 767–72 (2019).

[39] People by James v. Exxon Mobil Corp., 65 Misc. 3d 1233(A), 119 N.Y.S.3d 829 (N.Y. Sup. Ct. 2019). Massachusetts has also filed suit against Exxon Mobil Corp. raising similar claims under the Massachusetts Consumer Protection Act, including also a claim that Exxon engages in greenwashing of certain fossil fuel products.

[40] Ramirez v. Exxon Mobil Corp., 334 F. Supp. 3d 832 (N.D. Tex. 2018) (denying a motion to dismiss as to most named defendants).

and impacts will continue to support and strengthen private climate change governance.

CONCLUSION

As described in Chapter 5, one way that climate change litigation contributes to climate change policy is by helping to clarify what existing statutes, common law doctrines, and/or constitutional provisions require the government to do to mitigate climate change. Another important way that climate change litigation contributes to climate change policy is by ascertaining whether and how common law doctrines or federal securities laws apply to private climate change actors, such as large emitters of GHGs or producers of fossil fuels. Although the legal inquiries underlying efforts to apply law to governments and private interests are distinct (interpreting the scope of section 111(d) of the CAA versus analyzing what constitutes unreasonable interference for purposes of public nuisance, for example), developments in one area influence another. For example, the decision that the CAA applies to some emissions of GHGs resulted in the finding that federal common is, to some extent, displaced. Understanding how legal actions are defining understandings of both government and private responsibilities relating to climate change is thus important for an overall appreciation of how law is shaping our response to climate change.

8. Why the individual ethics of greenhouse gas emissions matters to climate law

Karl S. Coplan

I INTRODUCTION

This book is a primer on existing and potential responses of national and international legal systems to anthropogenic global warming. By definition, the response of legal systems is a collective, governmental response that seems to be independent of any considerations of individual ethics. Indeed, many activists for government response to climate change draw a sharp distinction between individual lifestyle greenhouse gas emissions and political legal measures, arguing that "systems change, not individual change" is needed.[1]

But individual lifestyle emissions of the wealthiest 10% of the world's population (which includes the median US household) account for well over half of global lifestyle emissions.[2]

Residents of the developed world make an outsized contribution to climate change through their choices to live in large, climate-controlled houses far from their places of employment, drive alone in inefficiently large fossil fuel-powered vehicles to get to work, consume diets high in red meat, and take far-flung vacations in jet aircraft that also make an outsized greenhouse gas contribution for each passenger. International and domestic legal mechanisms to address climate change are discussed in Chapters 1 and 2 of this book,

---

[1] Dr. Peter Critchley, The Climate Commitment: The Need for Common Agreement and Action on Climate:

Comments on the U.S. withdrawal from the Paris Climate Accord, available at https://www.academia.edu/33416412/The_Climate_Commitment_The_Need_for_Common_Agreement_and_Climate_Action_Comments_on_the_U_S_withdrawal_from_the_Paris_Climate_Accord (Sept., 2017).

[2] Oxfam, Extreme Carbon Inequality (Dec. 2015) (available at https://www-cdn.oxfam.org/s3fs-public/file_attachments/mb-extreme-carbon-inequality-021215-en.pdf).

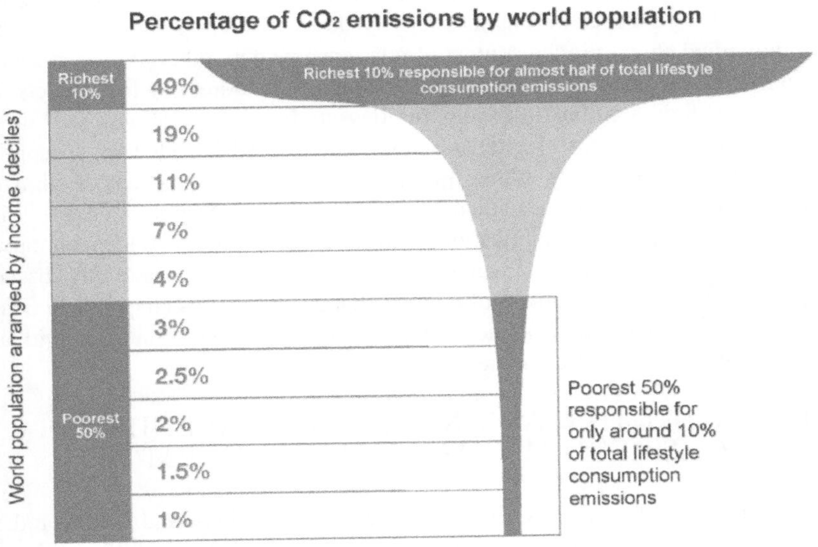

Percentage of CO₂ emissions by world population

Richest 10%	49%	Richest 10% responsible for almost half of total lifestyle consumption emissions
	19%	
	11%	
	7%	
	4%	
	3%	
	2.5%	
Poorest 50%	2%	Poorest 50% responsible for only around 10% of total lifestyle consumption emissions
	1.5%	
	1%	

World population arranged by income (deciles)

Figure 8.1   *Global income deciles and associated lifestyle consumption emissions*

respectively. However, legal mechanisms cannot address climate change without addressing this imbalance in individual emissions. "System change" versus "individual change" may thus be a false distinction, as a system of law and politics will only impose restraints on unfettered freedom of markets and individual choice when the restraints are perceived as driven by a moral imperative to respond to a problem. In this way, law does not direct individual morality so much as follow collective notions of individual morality. Many of the mitigation and compensation approaches discussed in this book depend on changing individual patterns of energy consumption, transportation, housing, dietary choices, and leisure travel. In order for society to accept these changes, there must be a consensus that these changes are the morally correct thing to do. This is equally true internationally as well as nationally, as the Paris Agreement on climate mitigation (discussed in Chapter 1) depends on voluntary, nationally determined, mitigation contributions. So, individual lifestyle emissions do matter, as widespread adoption of a greenhouse gas-conscious lifestyle in the developed world helps lay the political groundwork for an effective response.

Yet classical ethical systems do not easily address the problem of global climate change. Ethical thought is largely directed at assessing the morality of individual choices in the context of relationships with other individuals and communities. Ethics has not developed a context for addressing the impacts of collective choices on far-flung communities and future generations.

What follows in this chapter is a highly simplified introduction to application of some of the prevailing theories of ethics to the problem of climate change and the role of individual lifestyle consumption choices.[3] First, this chapter will consider the application of utilitarian ethics to lifestyle greenhouse gas emissions. Second, this chapter will consider the application of duty-based ethical systems to the problem of climate change. Third, this chapter will consider the application of principles of virtue ethics to the question of individual lifestyle contributions to climate change.

II   THE UTILITARIAN CASE FOR GREENHOUSE GAS REDUCTIONS: AVOIDING HARM

Utilitarian ethics considers the consequences of actions and assesses the morality of a given action based on the desirability or undesirability of these consequences. In its classic formulation, the desirability of an act's consequences turns on the net effect on aggregate human happiness.

A   Anthropocentric Utilitarianism

In its purest historical form, utilitarianism assesses the morality of an act based on its consequences for human welfare, measured in terms of the greatest happiness for the greatest number of people. Several philosophical writers have made the ethical case for limiting greenhouse gas emissions based on utilitarian ethics.[4] In essence, the primary argument for limiting—

---

[3]   For a more complete introduction, see James Garvey, The Ethics of Climate Change (Continuum International Publishing Group, London 2008); *see also* Stephen M. Gardiner, Ethics and Global Climate Change, *in* Climate Ethics: Essential Readings (Stephen M. Gardiner, Simon Caney, Dale Jamison, Henry Shue, eds., Oxford University Press, New York 2010).

[4]   *See, e.g.*, Paul Baer, *Equity, Greenhouse Gas Emissions, and Global Common Resources, in* CLIMATE CHANGE: A SURVEY 394 (Stephen Schneider et al. eds., 2002); John C. Dernbach & Donald A. Brown, *The Ethical Responsibility to Reduce Energy Consumption*, 37 HOFSTRA L. REV. 985 (2009); Richard C.J. Somerville, *The Ethics of Climate Change*, YALE ENV'T 360 (June 3, 2008), http://e360.yale.edu/feature/the _ethics_of_climate_change/1365/ [https://perma.cc/86QD-UE9N].

and ultimately ceasing—greenhouse gas emissions proceeds from the basic ethical principle of avoiding harm to others.[5] Burning fossil fuels results in greenhouse gas emissions that will change the global climate in ways that will interfere with food production and cause sea-level rise.[6] These changes will cause grievous harms to people globally, from food shortages, flooding, and civil strife as climate refugees move to higher ground.[7] Even if the global temperature increase is limited to 2°C above pre-industrial levels, serious environmental and human harms are likely to result: the complete inundation and destruction of low-lying island nations, and a high-risk of casualties affecting millions of people from extreme weather events including hurricanes, floods, and heat waves in sensitive areas.[8] The harms to low-lying island nations are likely to occur even if global warming is limited to the more ambitious 1.5°C aspirations of the Paris Agreement. As the global temperature increases, the risk of disruptions to unique ecosystems, extreme weather events, inequitable distribution of impacts burdening the poor, global aggregate impacts, and large-scale events all become "high" or "very high."[9] As these thresholds are exceeded, the partial loss of Arctic summer ice and the complete loss of the West Antarctic Ice Sheet become likely, causing even more extreme sea-level rise, coastal inundation, and climate refugees.[10] Additionally, the risk of reaching a global "tipping point" where thermal feedbacks cause a climate shift to a much hotter global climate unrecognizable to human beings is amplified.[11]

---

[5]   *See* Dernbach & Brown, *supra* note 27.

[6]   *Climate Impacts on Global Issues*, EPA, https://www3.epa.gov/climatechange/impacts/international.html [https://perma.cc/B8JF-2Z9U] (last visited Feb. 2, 2016).

[7]   *See id.*

[8]   *See* Petra Tschakert, Commentary, *1.5°C or 2°C: A Conduit's View from the Science-Policy Interface at COP20 in Lima, Peru*, Climate Change Responses, 2015 2:3, at 1, http://climatechangeresponses.biomedcentral.com/articles/10.1186/s40665-015-0010-z [https://perma.cc/EU5Q-VKML].

[9]   *Id.* at 9 fig.3; IPCC TOP LEVEL FINDINGS, *supra* note 3.

[10]   *See Global Warming Puts the Arctic on Thin Ice*, NAT. RES. DEF. COUNCIL, http://www.nrdc.org/globalwarming/qthinice.asp [https://perma.cc/5KBN-SXHZ] (last visited Mar. 8, 2016); *see also* Carolyn Gramling, *Just a Nudge Could Collapse West Antarctic Ice Sheet, Raise Sea Levels 3 Meters*, SCI. MAG. (Nov. 2, 2015, 3:00 PM), http://www.sciencemag.org/news/2015/11/just-nudge-could-collapse-west-antarctic-ice-sheet-raise-sea-levels-3-meters [https://perma.cc/R2CN-XVKC].

[11]   JOHN C. AYERS, GLOBAL CLIMATE CHANGE 13 (2011), http://www.vanderbilt.edu/Sustainability/book/S1C6.pdf [https://perma.cc/92BU-7GSF]; Brian Kahn, *Scientists Predict Huge Sea Level Rise Even If We Limit Climate Change*, GUARDIAN (July 10, 2015), http://www.theguardian.com/environment/2015/jul/10/scientists-predict-huge-sea-level-rise-even-if-we-limit-climate-change [https://perma.cc/R2TT-SNX5]; *see also* Anders Levermann & Johannes Feldmann, *Collapse of the West Antarctic Ice Sheet After Local Destabilization of the Amundsen Basin*, 112 PROC. NAT'L ACAD. SCI. 14191 (2015).

These substantial harms make it unethical to cause greenhouse gas emissions in excess of a nation's or an individual's fair share, and these harms will make it unethical to emit any greenhouse gases once the global increment is depleted.

Although some utilitarian ethical systems allow for harm to others to serve a greater good, no system of ethics allows grievous harms to others to provide luxuries to some. This has led some ethicists to conclude that greenhouse gas emissions, other than for basic sustenance and shelter, are unethical—at least as long as no allocation system exists to prevent the harms from occurring.[12] A white paper prepared by the Rock Institute has concluded that no system of ethics would justify continued emissions of greenhouse gases by developed nations in excess of an equal per capita allocation of the remaining GHG increment.[13] Principles of distributive justice require priority be given to the least well off, and principles of compensatory justice require that those who have benefited most from past greenhouse gas emissions have the weakest claim to an allocation exceeding equal distribution.[14]

On the other hand, some have argued that no ethical responsibility is owed to persons not yet in existence.[15] However, as Professor Simon Caney points out, this argument fails on two grounds.[16] First, climate change is harming and will harm people who already exist—today's children will suffer the future impacts of climate change.[17] Second, there is no ethically valid reason to ignore harms to future persons.[18] Future persons will have the same fundamental right to the basic human needs for food and shelter as people now in existence.[19]

Another possible objection to the existence of a utilitarian ethical duty to avoid greenhouse gas emissions is the collective nature of the harm: No single individual's or nation's greenhouse gas emissions can be independently responsible for the global climate harms.[20] As there is no direct causal link between any individual's use of fossil fuels and harm to any other person—present or future—the ethical case for limiting individual greenhouse gas

---

[12]   *See* Dernbach & Brown, *supra* note 27; Simon Caney, *Justice and the Distribution of Greenhouse Gas Emissions*, 5 J. GLOBAL ETHICS 125 (2009).

[13]   *See generally* DAVID BROWN ET AL., ROCK ETHICS INSTITUTE, WHITE PAPER ON THE ETHICAL DIMENSIONS OF CLIMATE CHANGE (2006), http://rockethics.psu.edu/ documents/whitepapers/edccwhitepaper.pdf [https://perma.cc/6DUP-8NG5].

[14]   *See id.*

[15]   *See* Simon Caney, *Cosmopolitan Justice, Rights and Global Climate Change*, 19 CAN. J.L. & JURIS. 255 (2006).

[16]   *See id.*

[17]   *Id.* at 272–73.

[18]   *See id.* at 263.

[19]   *See id.*

[20]   *See* Joakim Sandberg, *"My Emissions Make No Difference": Climate Change and the Argument from Inconsequentialism*, 33 ENVTL. ETHICS 229 (2011).

emissions is not as strong as consequentialist ethical proscriptions against homicide or assault, for example. However, there should be an ethical duty to avoid aggregate harms once it is clearly apparent that they are occurring.[21] This principle is similar to the imposition of joint and several liability for joint tortfeasors, where harm is not divisible.[22]

This causation argument has been refuted by other ethical writers. Professor Ben Amassi draws on the threshold nature of climate change causality, reasoning that since all individual emissions increase the likelihood that climate change thresholds will be surpassed, and since no individual can be sure that their own emissions will not cause the threshold to be passed, individual "luxury" emissions are morally wrong even without certainty of a but-for causal relationship to a climate change harm to any individual.[23] He also notes the "contagious" nature of luxury emissions, which encourages other people to engage in such conduct and expands the effective emissions beyond those of the individual actor.

Professor Amassi thus responds to the non-consequentialist argument against moral responsibility for climate impacts with an argument based on incremental risk. However, there is a more fundamental flaw in the non-consequentialist argument against moral responsibility for individual climate impacts: Its fundamental scientific premise is flawed. The non-consequentialist argument is based on the premise that climate impacts will occur once a threshold of aggregate carbon emissions has been passed, and that no individual can be said to have caused aggregate emissions to pass that threshold. In fact, according to the IPCC, global warming impacts are not threshold based, but largely linear: "Multiple lines of evidence indicate a strong, consistent, almost linear relationship between cumulative CO_2 emissions and projected global temperature change to the year 2100."[24] In fact all greenhouse gas emissions make the

---

[21] WOUTER PEETERS ET AL., CLIMATE CHANGE AND INDIVIDUAL RESPONSIBILITY (2015).

[22] Burlington N. & Santa Fe Ry. Co. v. United States, 556 U.S. 599 (2009); United States v. BestFoods Corp., 524 U.S. 51 (1998); United States v. Ne. Pharm. & Chem. Co., 810 F.2d 726 (8th Cir. 1986); United States v. Wade, 577 F. Supp. 1326 (E.D. Pa. 1983); Landers v. E. Tex. Salt Water Disposal Co., 248 S.W.2d 731 (Tex. 1952).

[23] Ben Almassi, *Climate Change and the Ethics of Individual Emissions: A Response to Sinnott-Armstrong*, 4 PERSP. INT'L POSTGRADUATE J. PHIL. 15 (2012). Almassi responded to Walter Sinnot-Armstrong's suggestion that, due to the lack of but-for causation for climate change, there is nothing morally wrong with driving a gas guzzling SUV "just for fun." *See id.*; *see also* Walter Sinnott-Armstrong, *It's Not My Fault: Global Warming and Individual Moral Obligations*, *in* PERSPECTIVES ON CLIMATE CHANGE 221–53 (W. Sinnott-Armstrong & R. Howarth eds., 2005). *See id.* at 16.

[24] IPCC, Climate Change 2014 Synthesis Report Summary for Policymakers at 8 (available at https://www.ipcc.ch/site/assets/uploads/2018/02/AR5_SYR_FINAL _SPM.pdf).

harms of global warming incrementally worse. Nor is the incremental harm associated with the emissions of one individual inconsequentially small. One writer has calculated that the lifetime emissions associated with the lifestyle of the average American will cause death or grievous harm to two persons in the future.[25]

B   Non-anthropocentric Utilitarian Ethics

It bears noting that the utilitarian concern for harms to persons—present and future—is not the only possible utilitarian ethical basis to limit or ban climate-altering greenhouse gas emissions. Utilitarian concerns solely based on the adverse effects of climate change on human beings constitute a form of anthropocentric ethics—one that only recognizes intrinsic value in human beings, but not in environmental integrity itself.[26] The environmental ethics awakening of the mid-twentieth century recognized that non-human animals and natural systems also have intrinsic value worthy of recognition in any system of utilitarian ethics. The so-called "deep ecology movement" recognized that all living things have their own intrinsic value.[27] Environmental ethics that respect natural systems and hold that humans have no ethical right to destroy them would hold that humans must avoid activities that destroy nature—including the natural climate system. This is the basis of Aldo Leopold's "land ethic": the idea that something is right when it tends to preserve the integrity, stability, and beauty of the biotic community, and wrong when it tends otherwise.[28]

This environmental ethic underlies such legislation as the Endangered Species Act and the Wilderness Act.[29] It has its roots in the basic human moral instincts of purity[30] and the corresponding natural human appreciation for

---

[25]   John Nolt, How Harmful Are the Average American's Greenhouse Gas Emissions?, Ethics, Policy and Environment, 14:1, 3–10 (2011).

[26]   Andrew Brennan & Yeuk-Sze Lo, *Environmental Ethics*, *in* STANFORD ENCYCLOPEDIA OF ETHICS (E. N. Zalta ed., 2015), http://plato.stanford.edu/entries/ethics-environmental/ [https://perma.cc/7QDN-28E9].

[27]   Arne Naess, *The Shallow and the Deep, Long Range Ecology Movement*, 16 INQUIRY 95 (1973), *reprinted in* THE DEEP ECOLOGY MOVEMENT: AN INTRODUCTORY ANTHOLOGY 3–9 (Alan Dregson & Yuichi Inoue eds., 1995).

[28]   ALDO LEOPOLD, A SAND COUNTY ALMANAC 262 (1949).

[29]   *See* Endangered Species Act of 1973, 16 U.S.C. §§ 1533–44 (2012); Wilderness Act, 16 U.S.C. §§ 1131–1136 (2012); *see also* Holly Doremus, *Restoring Endangered Species: The Importance of Being Wild*, 23 HARV. ENVTL. L. REV. 1, 13 (1999).

[30]   *See* JONATHAN HAIDT, THE RIGHTEOUS MIND: WHY GOOD PEOPLE ARE DIVIDED BY POLITICS AND RELIGION 15 (2012).

nature.[31] This basic principle of environmentalism supplements the anthropocentric utilitarian harm-avoidance arguments for banning fossil fuels.

III   DUTY-BASED NON-UTILITARIAN ETHICS

This environmental ethic is itself a form of utilitarianism, one that recognizes non-human ecological values as goal-worthy. Non-utilitarian approaches to ethics exist as well. Deontological ethics is a form of rule-based or authority-based ethics; actions are judged by their consistency or inconsistency with an ethical maxim establishing ethical duties.[32] The Ten Commandments are a form of deontological ethics.[33] Categorical imperatives, such as the Golden Rule, positing that one should do unto others as one would have others do unto oneself, are also a form of deontological ethics. Aldo Leopold's land ethic might also be considered a form of deontological ethics.[34] Application of deontological ethics to the issue of climate change may be problematic, as deontological ethics are indeterminate—there is no objective way to determine whose "authoritative" ethical maxims are the correct ones. The authoritative nature of any particular ethical maxim depends on a sort of social and political consensus.

With this caveat that deontological—authoritarian, rule-based—ethics may be a weak source of legal proscriptions, it bears noting that world religious leaders are moving in the direction of recognizing a proscription against harming the global climate system on theological grounds. The most striking recent development, of course, is Pope Francis's *Laudato Si'*—an encyclical on the environment and climate change issued on June 18, 2015.[35] Pope Francis wrote that:

> The creation accounts in the book of Genesis contain, in their own symbolic and narrative language, profound teachings about human existence and its historical

---

[31]   Nicholas A. Robinson, *Evolved Norms: A Canon for the Anthropocene, in* RULE OF LAW FOR NATURE 46–71 (Christina Voigt ed., 2014).

[32]   *See* J.P. Moreland, *Ethics Theories: Utilitarianism vs. Deontological Ethics*, CHRISTIAN RES. INST. (Apr. 17, 2009), http://www.equip.org/article/ethics-theories-utilitarianism-vs-deontological-ethics/ [https://perma.cc/HS3Q-92Q7].

[33]   *See* WILFRED BECKERMAN, ECONOMICS AS APPLIED ETHICS: VALUE JUDGEMENTS IN WELFARE ECONOMICS 80 (2011).

[34]   J. BAIRD CALLICOTT, IN DEFENSE OF THE LAND ETHIC: ESSAYS IN ENVIRONMENTAL PHILOSOPHY (1989).

[35]   *See* POPE FRANCIS, ENCYCLICAL LETTER LAUDATO SI' OF THE HOLY FATHER FRANCIS ON CARE OF OUR COMMON HOME (June 18, 2015), http://w2.vatican.va/content/dam/   francesco/pdf/encyclicals/documents/papa-francesco_20150524_enciclica-laudato-si_en.pdf [https://perma.cc/LJQ9-R7ZD].

reality. They suggest that human life is grounded in three fundamental and closely intertwined relationships: with God, with our neighbor and with the earth itself.[36] A very solid scientific consensus indicates that we are presently witnessing a disturbing warming of the climatic system. In recent decades this warming has been accompanied by a constant rise in the sea level and, it would appear, by an increase of extreme weather events, even if a scientifically determinable cause cannot be assigned to each particular phenomenon. Humanity is called to recognize the need for changes of lifestyle, production and consumption, in order to combat this warming or at least the human causes which produce or aggravate it.[37]

Pope Francis has thus brought Catholic orthodoxy into the creation care movement of Christianity, which similarly holds that humans are stewards of God's creation and have a responsibility to preserve the global ecosystem intact.[38] This recognition of stewardship obligations stands in contrast to the so-called "dominion" theory of Judeo-Christian environmental thought, which relies on the Biblical injunction to humans to "be fruitful and multiply, and replenish the earth, and subdue it: and have dominion over the fish of the sea, and over fowl of the air, and over every living thing that moveth upon the earth."[39] Following Pope Francis's lead, a group of Islamic religious and environmental leaders issued its own call to phase out non-renewable energy and stop greenhouse gas emissions no later than 2050.[40] There is thus some deontological ethical support for a prohibition on burning fossil fuels.

IV   VIRTUE ETHICS AND INDIVIDUAL CLIMATE EMISSIONS

The third major rule of ethical thought—virtue ethics—also provides ambiguous support for a prohibition against fossil fuel use. Virtue ethics looks to the motivations and character traits of the moral actor rather than to the effects (utilitarian) or rule-compliance (deontology) of the acts in question.[41] Virtue

---

[36]   *Id.* at para. 66.

[37]   *Id.* at para. 23.

[38]   *See European Parliament Address by Pope Francis*, C-SPAN (Nov. 25, 2014), http://www.c-span.org/video/?323063-1/pope-francis-address-european-parliament [http://perma.cc/24W6-8GU2].

[39]   *See* ANDREW KERNOHAN, ENVIRONMENTAL ETHICS: AN INTERACTIVE INTRODUCTION 194 (2012).

[40]   Denise Hassanzade Ajiri, *Islamic Leaders Echo Pope's Call for Action on Climate Change*, CHRISTIAN SCI. MONITOR (Aug. 18, 2015), http://www.csmonitor.com/Science/2015/0818/   Islamic-leaders-echo-pope-s-call-for-action-on-climate-change [https://perma.cc/U6NX-66XZ].

[41]   *See* P. Gardiner, *A Virtue Ethics Approach to Moral Dilemmas in Medicine*, 29 J. MED. ETHICS 297 (2003).

ethics has its roots in the ancient Greek philosophical traditions of Socrates and Aristotle, and seeks to promote a state of individual harmony through actions consistent with personal virtues such as courage, honesty, rationality, friendliness, and loyalty.[42]

As virtue ethics seeks to promote individual self-realization and a state of harmony, its application to a problem such as climate change may seem impossibly anthropocentric and subjective. However, several ethical writers have suggested ways that a system of virtue ethics is at least consistent with, and perhaps demands, attention to one's individual contribution to climate change. Some virtue ethicists have suggested that the virtue ethics values of love, respect, and care may apply to the non-human environment as well.[43] Aristotelian virtue ethics stresses deliberation and mindfulness in all individual actions, leading one modern Aristotelian to conclude that "Aristotle would have been appalled at the mess we have made of our world by a failure to take our responsibility to it and its non-human residents seriously."[44]

Aristotelian virtue ethics stresses the relationship between deliberately living in accord with virtuous character traits and the overall well-being and flourishing of the individual actor. Choosing to moderate consumption and be mindful of one's contribution to the catastrophic impacts of climate change is thus choiceworthy not because reducing consumption can be proven to reduce specific impacts, but because a life lived consistently with an understanding of the individual in contributing to climate change will be a less dissonant and more contented life. Ethics Professor Dale Jamieson makes the argument for a virtue ethics approach in his essay "When Utilitarians Should be Virtue Theorists":

> Nor is it an "all or nothing" phenomenon. Collectively, we can prevent or mitigate various aspects of global environmental change, and an individual agent can affect collective behavior in several ways. One's behavior in producing and consuming is important for its immediate environmental impacts, and also for the example-setting and role modeling dimensions of the behavior (...) Nor does an environmentally friendly lifestyle have to be a miserable one. Even if in the end one's values do not prevail, there is comfort and satisfaction in living in accordance with one's ideals.[45]

---

[42] *See id.*
[43] *See* John O'Neill, *The Varieties of Intrinsic Value*, 75 MONIST 119 (1992); *see also* JOHN BARRY, RETHINKING GREEN POLITICS 32 (1999).
[44] Edith Hall, Aristotle's Way: How Ancient Wisdom Can Change Your Life 170 (Penguin Press, New York 2018).
[45] Dale Jamieson, When Utilitarians Should Be Virtue Theorists, in Climate Ethics, supra n. 3, at 324.

CONCLUSION

An effective collective response to climate change, in the form of legal regimes promoting mitigation of emissions, may well depend on collective acknowledgment of a shared ethical responsibility for consumption-related greenhouse gas emissions. Although far from settled, the prevailing ethical theories of utilitarianism, deontology, and virtue ethics lend support to the idea that individuals have a moral responsibility to reduce and moderate their lifestyle greenhouse gas emissions. Individual consumption choices matter ethically, based both on the principle of avoidance of harm to people and on the ethic of non-interference with natural ecosystems. Recognition of the duty of stewardship for the natural ecosystems of the Earth likewise supports this climate ethic. An individual climate ethic is also consistent with the virtues of moderation and mindfulness, and of promoting lifestyles that are harmonious with recognition of the threat of climate change and concern for avoiding contributing to those threats.

Index